The Clinical ORTHOPEDIC ASSESSMENT *Guide*

Janice K. Loudon, PhD, PT, SCS, ATC
University of Kansas Medical Center

Stephania L. Bell, MS, PT
University of Kansas Medical Center

Jane M. Johnston, MS, PT
Kansas City Physical Therapy Group

Human Kinetics

Library of Congress Cataloging-In-Publication Data

Loudon, Janice, 1959-
    The clinical orthopedic assessment guide / Janice Loudon,
Stephania Bell, Jane Johnston.
        p. cm.
    Includes bibliographical references and index.
    ISBN: 0-88011-507-6
    1. Orthopedics--Diagnosis--Handbooks, manuals, etc.   I. Bell,
Stephania, 1966- . II. Johnston, Jane (Jane M.), 1943-
III. Title.
    [DNLM: 1. Musculoskeletal Diseases--diagnosis--handbooks.
2. Joints--physiology--handbooks. 3. Joints--physiopathology--handbooks.
4. Physical Examination--handbooks. 5. Orthopedics--handbooks. WE 39 L886c 1998]
    RD734.L68   1998
    616.7'075--dc21
    DNLM/DLC
    for Library of Congress
                                                 97-49234
                                                    CIP

ISBN: 0-88011-507-6

Acquisitions Editors: Rik Washburn and Loarn Robertson; **Developmental Editors:** Julia Anderson and Julie Rhoda; **Assistant Editors:** Sandra Merz Bott and Cassandra Mitchell; **Editorial Assistants:** Jennifer Jeanne Hemphill and Laura T. Seversen; **Copyeditor:** Joyce Sexton; **Proofreader:** Debra Aglaia; **Indexer:** Sandi Schroeder; **Graphic Designer:** Fred Starbird; **Graphic Artist:** Joe Bellis; **Cover Designer:** Jack Davis; **Cover Illustrator:** Kristin Mount; **Illustrators:** Kristin Mount, Tim Offenstein, and Joe Bellis; **Printer:** United Graphics

Printed in the United States of America   10 9 8 7 6 5 4 3 2 1

Human Kinetics
Web site: http://www.humankinetics.com/

United States: Human Kinetics, P.O. Box 5076, Champaign, IL 61825-5076
1-800-747-4457
e-mail: humank@hkusa.com

Canada: Human Kinetics, Box 24040, Windsor, ON N8Y 4Y9
1-800-465-7301 (in Canada only)
e-mail: humank@hkcanada.com

Europe: Human Kinetics, P.O. Box IW14, Leeds LS16 6TR, United Kingdom
(44) 1132 781708
e-mail: humank@hkeurope.com

Australia: Human Kinetics, 57A Price Avenue, Lower Mitcham, South Australia 5062
(088) 277 1555
e-mail: humank@hkaustralia.com

New Zealand: Human Kinetics, P.O. Box 105-231, Auckland 1
(09) 523 3462
e-mail: humank@hknewz.com

CONTENTS

CREDITS

Figures 3.4b, 8.6a, 9.4, 9.5, 9.6a, 9.7, 9.12a, 9.13, 11.6, 12.8, and 16.1; Tables 3.2, 4.3, 5.2, 6.3, 7.6, 8.4, 9.5, 11.4, 12.4, 12.8, and 13.5: D.J. Magee, 1992, *Orthopedic physical assessment.* (Philadelphia: Saunders).

Tables 3.1, 4.2, 5.1, 6.2, 7.5, 8.3, 9.4, 11.3, 12.3, and 13.4: J. Hamill and K. Knutzen, 1995, *Biomechanical basis of human movement.* (Baltimore: Williams & Wilkins).

Tables 3.3, 6.4, 7.7, 8.5, 9.6, 11.5, 12.5, and 13.6: C. Wadsworth, 1988, *Manual examination and treatment of the spine and extremities.* (Baltimore: Williams & Wilkins).

Figure 5.8: W. Haymaker and B. Woodall, Eds., 1953, *Peripheral nerve injuries.* 2nd ed. (Philadelphia: Saunders).

Figures 7.2, 7.3, 7.16, 7.17, 8.2, 8.3, 8.5, 9.1, 9.2, 9.8, 9.10, 9.11, 9.14, 11.1, 11.2, 12.1, 12.9, 13.1, 13.2, 13.3, 13.5, 13.6, 13.7, 13.9, and 13.10: C. Kisner and L. Colby, 1996, *Therapeutic exercise: Foundations and techniques,* 3rd ed. (Philadelphia: Davis).

Figures 7.15, 9.16, 10.1, 11.5, and 11.11: A. Hartley, 1990, *Practical joint assessment: A sports medicine manual.* (St. Louis: Mosby).

Tables 10.1, 10.3, 10.4, 14.1, and 14.5: M.L. Palmer and M. Epler, 1990, *Clinical assessment procedures in physical therapy.* (Philadelphia: Lippincott).

Table 14.2: W.H. Harris, 1969, "Traumatic arthritis of the hip after dislocation and acetabular fracture: Treatment by mold arthroplasty," *Journal of Bone and Joint Surgery* 51: 737-755.

Table 14.3: F.R. Noyes, et al., 1984, "Functional disability in the anterior cruciate insufficient knee syndrome," *Sports Medicine* 1: 286-287.

Table 14.6: D. Seligson, J. Sassman, and M. Pope, 1980, "Ankle instability: Evaluation of lateral ligaments," *American Journal of Sports Medicine* 8: 39.

Table 14.7: D.B. Reuben, 1990, "An objective measure of physical function of elderly outpatients: The Physical Performance Test," *Journal of the American Geriatrics Society* 38 (10): 1111.

Table 14.8: Boston University School of Public Health 1990.

Tables 15.1 and 15.2: F.P. Kendall, 1993, *Muscles: Testing and function,* 4th ed. (Baltimore: Williams & Wilkins).

Table 16.1: C. Norkin, 1994, Gait analysis. In *Physical rehabilitation: Assessment and treatment,* edited by S.B. O'Sullivan and T.J. Schmitz (Philadelphia: Davis), 177-179.

Figure 17.2: C. Norkin and P. Levangie, 1992, *Joint structure and function.* (Philadelphia: Davis), 482.

INTRODUCTION

This book provides a handy and inexpensive reference for orthopedic clinicians utilizing manual therapy techniques in patient assessment. The text is unique in that it presents information in a concise and easy-to-find manner with illustrations throughout to facilitate understanding. The text has been organized to make the same categories of information pertaining to each joint easy to find.

The book is divided into five parts. Part I introduces the text, covers basic arthrokinematic and osteokinematic principles, and operationally defines terminology used throughout *The Clinical Orthopedic Assessment Guide*. Part II contains head and trunk information, specifically about the temporomandibular joint (TMJ), cervical, thoracic, and lumbar spine, and the sacroiliac joints. Part III comprises the upper extremity and includes individual sections on the shoulder, the elbow and forearm, and the wrist and hand. Part IV covers the lower-extremity material on the hip, knee, and ankle and foot. Part V provides a synopsis of posture as well as walking and running gait. Both normal and abnormal posture and gait are described in order to facilitate clinicians' decision making for appropriate treatment. We also have included an appendix that lists the action and side effects of medications commonly taken by patients in or-

thopedic settings. Sections on specific upper-extremity and lower-extremity joints each focus on one anatomical body region and follow a similar format. First, we describe the type of joint and the articulating surfaces. Next we list degrees of freedom, active range of motion (if applicable), accessory movements, end-feel, capsular pattern, close- and loose-packed position, stability, and special tests. The remainder of the section describes arthrokinematics, neurological assessment, surface palpation, muscle origins and insertions, and actions and innervations. Finally, we have included tables at the end of most of these sections to summarize differential diagnoses.

Both the upper-extremity and lower-extremity portions of the book end with sections presenting batteries of functional tests. With the need for functional measures in a clinical setting, these two sections (sections 10 and 14) are intended to supplement a good orthopedic assessment. The functional tests listed are proven tests that are valid and reliable.

Part V includes numerous tables that describe posture and gait from head to toe. The phases of gait described by Rancho Los Amigos are presented in a condensed form for easy access.

PART

1

INTRODUCTION TO BIOMECHANICAL PRINCIPLES

SECTION 1

CLASSIFICATIONS AND DEFINITIONS

The basis for good orthopedic treatment is a good orthopedic assessment. The skilled orthopedic clinician must have a solid working knowledge of joint biomechanics in order to develop an appropriate treatment plan. This section defines the bases of biomechanics including articulation, type of joint, degrees of freedom, active range of motion, and accessory movement, end-feel, capsular pattern, close- and loose-packed positions, stability, and special tests (arthrokinematics). The section provides a review of terminology and serves as a layout for the rest of the guide. Further detailed information on arthrokinematics can be found in the bibliography.

JOINT BASICS

Articulation

The segment on articulation in each section of this book is used to describe joint shape and define the actual contact area between the bones that make up the joint articulation.

Type of Joint

Many forms of joint classifications can be found in the literature. This text classifies the joints as presented in *Cunningham's Textbook of Anatomy*. There are three classifications of joints, consisting of fibrous joints, cartilaginous joints, and synovial joints. The description in this subsection of the 12 types of joints begins with the joints that are the least mobile and progresses to the joints with the most motion.

Fibrous Joint (Synarthrosis)

Suture   A fibrous joint, found only in the skull, with minimal to no movement (figure 1.1).

Syndesmosis   Two bones connected by fibrous tissue that allows minimal movement; much denser than a suture.

Example: membranous connection (interosseous) between tibia and fibula (figure 1.2).

Gomphosis   This fibrous joint is analogous to a "peg" fitting into a socket, resulting in minimal movement.

Example: tooth in socket.

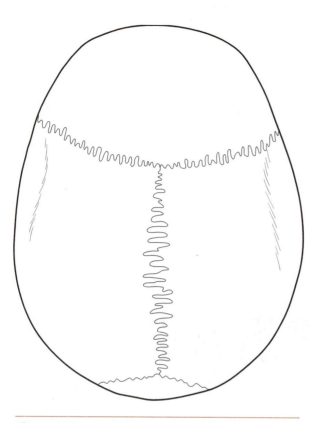

Figure 1.1 A suture in the skull.

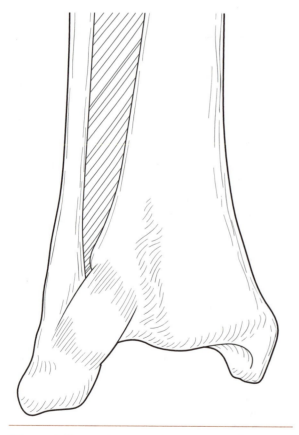

Figure 1.2 Syndesmosis between the tibia and fibula.

Cartilaginous Joint (Amphiarthrosis)

Synchondrosis Cartilaginous connection between two bones that eventually ossifies with maturity; virtually immobile.

Example: epiphyseal plate (figure 1.3).

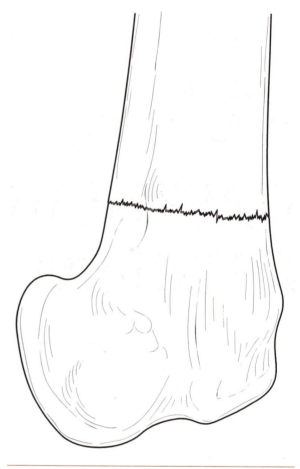

Figure 1.3 Synchondrosis of the epiphyseal plate.

Symphysis Joint consisting of two bones connected by hyaline cartilage and fibrocartilage; slightly movable.

Example: pubic symphysis (figure 1.4).

Synovial Joint (Diarthrosis)

Plane Gliding joint with opposing surfaces relatively flat.

Example: superior tibiofibular joint (figure 1.5).

Sellar (Saddle) Two bones each with reciprocal concavoconvex articular surfaces that fit together like a puzzle; biaxial; produces flexion and extension, abduction and adduction.

Example: carpometacarpal of thumb (figure 1.6).

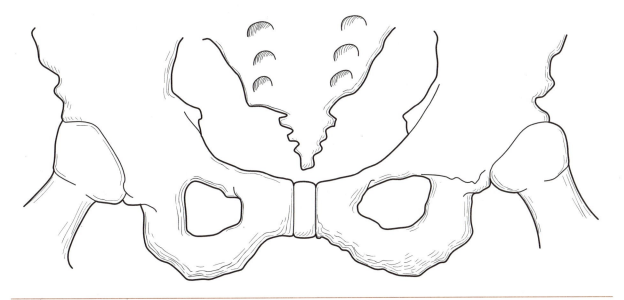

Figure 1.4 Pubic symphysis.

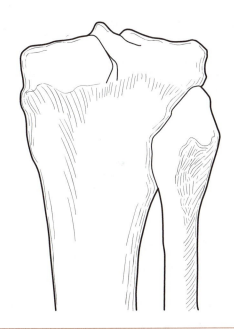

Figure 1.5 Plane of the superior tibiofibular joint.

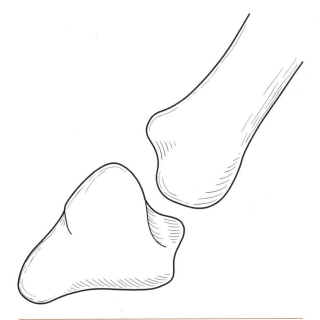

Figure 1.6 Sellar joint of the carpometacarpal of the thumb.

Ginglymus (Hinge) Two bones that articulate and permit motion in only one plane; uniaxial; flexion and extension.

Example: ulnohumeral joint (figure 1.7).

Trochoid (Pivot) One articulating surface cylindrical in shape rotates within a ring formed by bone and/or ligament, allows supination, pronation, and rotation.

Example: atlantoaxial joint (figure 1.8).

Spheroid An articulation in which one bone is convex (ball shaped) and rotates about a concave surface (socket) of a second bone; triaxial; all joint movements.

Example: glenohumeral joint (figure 1.9).

Condyloid "Ball and socket" type, but ligamentous constraints prevent rotation about a vertical axis.

Example: metacarpophalangeal joints of digits 2 to 5.

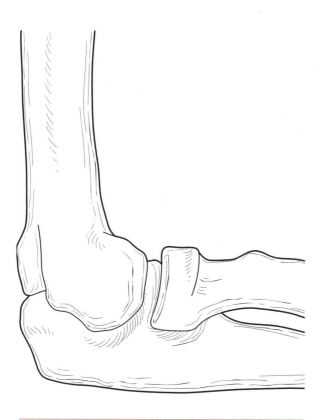

Figure 1.7 Ginglymus of the ulnohumeral joint.

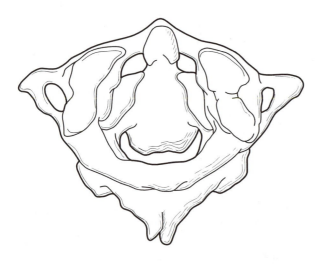

Figure 1.8 Trochoid of the atlantoaxial joint.

Ellipsoid Modified "ball and socket" in which one articulating surface is ellipsoid instead of spheroidal; biaxial; flexion and extension, abduction and adduction.

Example: radiocarpal joint.

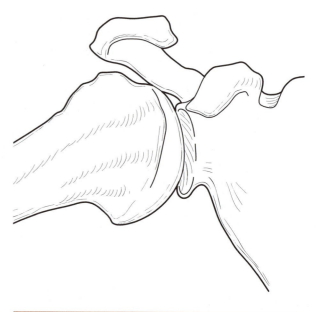

Figure 1.9 Spheroid of the glenohumeral joint.

Degrees of Freedom

Degrees of freedom refer to the number of movements that occur at a specific joint. For example, the knee has six degrees of freedom: rotational and linear motion occuring in three planes (coronal, sagittal, and tranverse).

Rotation

Flexion and Extension Occurs in a sagittal plane about a coronal axis (figure 1.10).

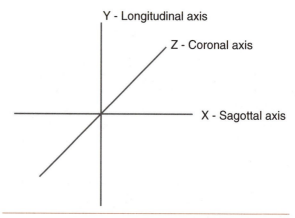

Y - Longitudinal axis

Z - Coronal axis

X - Sagottal axis

Figure 1.10 Rotation movement patterns.

Adduction and Abduction Occurs in a coronal plane about a sagittal axis.

Internal and External Rotation Occurs in a transverse plane about a longitudinal axis.

Translation

Anterior and Posterior   Occurs about a sagittal axis; linear movement volar and dorsal.

Lateral and Medial   Occurs about a coronal axis; linear movement side to side.

Superior and Inferior   Occurs about a longitudinal axis; linear movement cranial and caudal.

For purposes of this text, the degrees of freedom will be described for rotational motion only. Translation is described as a glide and is included below in the material on accessory movement

Axis of Motion

A joint's axis of motion is the axis about which it rotates.

Active Range of Motion

Active range of motion is the normal amount of motion that occurs in a joint in one of the three cardinal planes. It is measured in degrees and is dependent on the type of joint.

Accessory Movement

Roll

One portion of the joint surface rolls on another

- Surfaces are incongruent.
- New points on one surface contact new points at similar intervals on opposing surface.
- Rolling is always in same direction as angular movement.

Example: The distal femur rolls posteriorly during knee flexion.

Slide

One bone slides across another

- Surfaces are congruent.
- Same point on moving joint surface contacts various points on opposing surface.
- Sliding occurs in the opposite direction of angular movement if the moving joint surface is convex; and occurs in the same direction as angular movement if the moving surface is concave (most joint surfaces move with a combination of roll and slide).

Example: The proximal tibia slides anteriorly during knee extension.

Spin

One bone rotates perpendicular to joint plane of stationary bone

- Rotation occurs about a stationary mechanical axis.
- Usually occurs in conjunction with rolling and sliding.

Example: The proximal radius spins posteriorly on the radial notch with pronation.

Traction

Separation of two articulating surfaces

Distraction

Separation of two articulating surfaces perpendicular to axis of motion

Compression

Approximation of two articulating surfaces

CONCAVE-CONVEX RULE (KALTENBORN)

When a concave surface moves on a convex surface, roll and slide occur in the same direction. For example, in nonweight bearing knee extension, the tibia rolls and slides anteriorly on the fixed femur. When a convex surface moves on a concave surface, roll and slide are in the opposite direction. For example, in weight bearing knee extension, the femur rolls anterior and slides posterior on a fixed tibia. This is the primary principle by which joint mobilization treatment is determined (figures 1.11 and 1.12).

(continued)

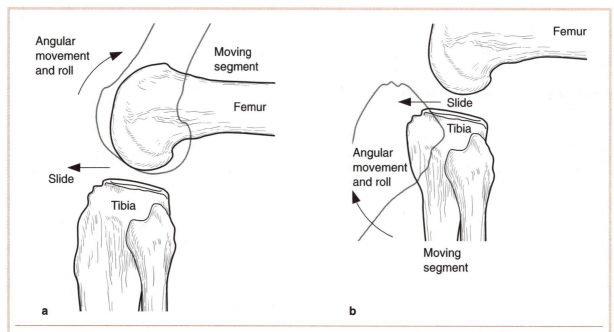

Figure 1.11 Joint mobilization: (*a*) movement of the femur on the tibia, and (*b*) movement of the tibia on the femur.

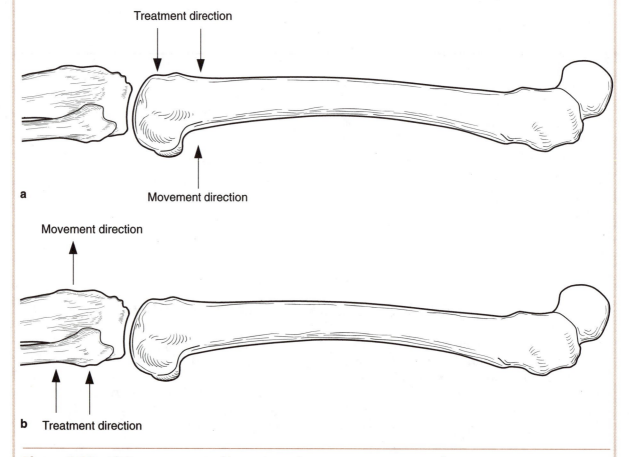

Figure 1.12 Gliding movement of bone depends on concave-convex surface.

Treatment Direction

The direction of treatment for accessory motion is generally dictated by the concave-convex rule. The operator moves the convex bone in the opposite direction as the restriction. The concave bone glides in the same direction as the restriction.

Contraindications to Joint Mobilization

1. Hypermobility and instability
2. Joint effusion
3. Infection and inflammation (acute)
4. Rheumatoid arthritis (active inflammatory stage)
5. Cancer (active metastasis)
6. Disturbed bowel or bladder function and perineal anesthesia: indicative of cauda equina compression
7. Cord compression: pins and needles sensation in bilateral hands and/or feet

Precautions to Joint Mobilization

1. Osteoporosis
2. Spondylolisthesis
3. Pregnancy
4. Positive vertebral artery test
5. Spinal fractures
6. Spinal surgery
7. Spinal stenosis
8. Neurological changes: pain associated with disturbances of reflexes, muscle power, or sensation

End-Feel: Physiological Limit of a Joint

Normal

Soft Give of end-point due to soft tissue approximation.
Example: elbow flexion.

Firm (Capsular) Hard or firm type of movement with some give.
Example: external rotation of shoulder.

Hard (Bony) Hard unyielding sensation.
Example: elbow extension.

Abnormal

Spasm Sudden dramatic arrest of movement accompanied by pain.
Example: inflamed joint.

Springy Similar to firm end-feel, but in range where one would not expect to find end-feel.
Example: torn meniscus of knee.

Empty The end-feel when the patient is in considerable pain with movement and will not allow you to continue passive range. No mechanical resistance is encountered.
Examples: subacute bursitis, neoplasm.

Firm (Capsular) A firm end-feel when normally the end-feel is soft or hard.
Example: synovitis or soft tissue edema.

Hard Abrupt halt in range at an abnormal range.
Examples: fracture, arthropathy.

PAIN-RESISTANCE SEQUENCE (CYRIAX)

1. *Pain before resistance* indicates an acute lesion. During performance of passive range of motion, pain occurs before the clinician meets resistance.
2. *Pain with resistance* indicates a subacute lesion. Pain occurs at the same time as resistance.
3. *Pain after resistance* indicates chronic lesion. The clinician notes resistance before the patient notes pain.

Capsular Pattern

The capsule is constructed of fibrous connective tissue and helps to maintain the integrity of synovial joints. A capsular pattern is a proportional limitation of movement in a joint that usually indicates injury or inflammation of the capsule. Tightness is noted within the joint capsule and ligaments. Each synovial joint has a characteristic capsular pattern. A noncapsular pattern, on the other hand, is limitation in a joint that is not characteristic of the capsular pattern. Possible causes include internal derangement and bursitis.

Close-Packed Position

Close-packed position usually refers to the extreme end of range of motion. The joint surfaces are maximally congruent, the ligaments and capsules are taut, and the joint possesses its greatest stability.

Loose-Packed Position

In this text, loose-packed denotes maximal loose-packed position. In loose-packed position the articular surfaces are relatively free to move in relation to one another. In this position the largest amount of joint play occurs. Ligaments and capsules are lax and dynamically safe.

Stability

Each joint has soft tissue structures that help to maintain the stability of that joint. These structures include the capsule, ligaments, and menisci. The subsection on stability defines the major stabilizing structures of each joint.

Special Tests

The subsection on special tests presents common special tests for each joint. Special tests are performed to differentially diagnose tissues that may be pathological or injured.

ARTHROKINEMATICS

The subsection on arthrokinematics summarizes the accessory and conjunct motion that accompanies each plane of motion for that joint. This information can help the clinician decide what mobilization technique is indicated to improve joint mobility.

NEUROLOGY

The neurology subsections present dermatomal, myotomal, and peripheral nerve information for each joint.

SURFACE PALPATION

The subsection on surface palpation is designed to give the clinician a quick guide to common palpatory areas for each joint.

MUSCLE ORIGIN AND INSERTION

The subsection on muscle origin and insertion lists the origin point and insertion point for muscles surrounding each joint.

MUSCLE ACTION AND INNERVATION

The subsection on muscle action and innervation lists muscle action about each joint and the nerve innervation for each muscle.

DIFFERENTIAL DIAGNOSIS

A table at the end of sections 3, 6-9, and 11-13 identifies common musculoskeletal diagnoses. This table is intended to help the clinician differentiate the tissue at fault. The table includes the name of the disorder, a description, onset of injury, symptoms, signs, and special tests.

SUBJECTIVE EXAMINATION AND ASSESSMENT

Although the majority of this text is directed at specific objective tests and assessments, we feel it is imperative that a detailed subjective exam or patient interview be performed at the outset of evaluation. Geoffrey Maitland, an Australian physical therapist who has contributed volumes to the world of orthopedic physical therapy, highlights the importance of thorough interrogation of the patient at the outset of treatment. A complete subjective exam should help the practitioner formulate a hypothesis regarding a patient's problem that can then be proven or altered accordingly as objective testing and treatment proceed. Thus, the information obtained during the subjective exam should guide the therapist's selection of objective tests and treatments, as well as the vigor with which they are carried out. The subjective examination or patient interview is critical in enabling the therapist to accurately assess the patient's problem. Additionally, a detailed patient interview offers an opportunity to establish a solid patient-therapist rapport. The patient learns that his or her information is of great value to the therapist who in turn generally encourages the patient to participate in resolving the problem.

Maitland emphasizes good communication between therapist and patient when obtaining information. Therefore it is essential to know not only which questions to ask, but why you are asking those questions. In other words, predetermine what is to be gained by proceeding with a certain line of questioning. This forethought allows you to determine whether the patient is providing the appropriate information and, if not, what strategies you can use to obtain the necessary information. For instance, if you ask the patient where her symptoms are located and she responds "in my leg," this is not particularly helpful in identifying what structure is likely to be at fault. However, you can follow up by clarifying: "Could you outline with your hands exactly where in your leg the symptoms are?" The patient then, in outlining the specific area of symptoms, may point you to a specific dermatome versus a peripheral nerve pattern. This information offers a small but important step in assisting you in differential diagnosis, and consequently, appropriate treatment.

The following guidelines summarize the types of questions Maitland recommends be asked of all patients. For a more in-depth explanation as to the relevance of these questions and for further reading on enhancing communication skills, refer to Maitland's *Vertebral Manipulation* (1986).

☐ Patient Profile

This refers to the patient's personal information such as age, occupation, current daily activity level (as compared to normal, if different), recreation or hobbies, and psychosocial factors such as other dependents, litigation, or workman's compensation. It is important to note the specifics of occupation and activities in order to gain an appreciation for what positions or movements the person must undergo on a daily basis. For instance, if a patient states that his occupation is office worker, this may entail a variety of things. Does he sit at a computer, talk on the phone, read, write, etc.? If the job entails multiple activities, how long does he engage in each one? It is easy to see that *office work* might mean different things to different people and that this may impact the therapist's choice of treatment and development of appropriate goals

☐ Symptom Description— Location and Type

The second step is to identify the precise location of symptoms. Ask the patient to outline with his or her hands (one finger if possible) exactly where the symptoms manifest themselves. Follow by retracing the symptom outline on the patient's body to verify the location. Then ask the patient to describe in his or her own words how the symptoms feel. It is important to continue to use the patient's words when referring to the symptoms. Not only is this more accurate for the therapist in that it encourages thinking in more correct terms, but it also conveys to the patient that the therapist was listening to the patient's original description. If the patient does not volunteer any of the following information when describing his or her symptoms, then be sure to determine whether the symptoms are deep or superficial, constant or intermittent, sharp or dull, variable or nonvariable, and whether any numbness or tingling is present. All of these factors may provide insight into the nature of the patient's condition and what structure or structures are at fault. Precise description of symptom location and type often provides an early clue for the therapist as to the structure(s) responsible for the patient's complaints. As a final step, be sure to verify that there are no complaints in the joints above or below the symptom area(s) all along the kinetic chain. For any upper-extremity complaint, check the wrist and hand, the elbow, the shoulder, and the neck. For any lower-extremity complaint, be sure to clear the foot and ankle, the knee, the hip, and the lumbar spine.

☐ Behavior of Symptoms

Now establish how the symptoms behave. If there is more than one area of symptoms, is there a relationship between the areas? For instance, if a patient complains of back pain and leg pain, does she describe a scenario in which the back pain worsens and then the leg pain worsens, or do the two behave totally independently of one another? Also determine what activities, movements, or positions aggravate the patient's symptoms to help you assess the severity and irritability of the patient's condition. Ask the patient what frequency or what duration of an aggravating activity is required to reproduce symptoms, to what extent they are reproduced, and then how long after cessation of the aggravating activity the symptoms disappear. Additionally, determine factors that ease the symptoms such as movements, positions, medications, etc. Also, determine the behavior of the patient's symptoms over a 24-hour period. Do the patient's symptoms wake her from sleep? If so, how long until she is able to return to sleep? Symptoms that wake a patient from sleep suggest a more severe condition. How are the symptoms first thing in the morning? How do the symptoms behave throughout the day? Again, answers to these questions may provide clues as to the nature as well as the severity of the patient's problem.

☐ Special Questions

In addition to interviewing the patient regarding his or her specific complaint, ask several questions to help rule out any contraindications to treatment or identifying factors that warrant caution in physical evaluation and treatment of the patient. We recommend asking the following questions of all patients presenting with any complaint of lumbar pain or extremity pain of unknown etiology.

1. How is your general health? (This informs the therapist of other ongoing medical processes and provides insight as to the patient's impression of his or her own health.)

2. Have you experienced any recent unexplained weight loss or weight gain? (This may point to possible systemic disease or tumor.)

3. Are you taking any medications currently? (Again, this gives information about general health and potential exercise precautions.)

4. Do you have any history of long-term steroid or anticoagulant use? (Either of these situations might present a precaution to joint mobilization because of increased risk of tissue damage or bleeding.)

5. Have you had any special tests such as x-ray, CT, MRI, etc., for this problem? If yes, are you aware of the results? (This may provide information for the therapist if the results are not present in the patient's chart. It may also be that the results are available but the patient is unaware of what they are.)

6. Have you had any trouble initiating urination or controlling bowels? Are you experiencing any numbness in the saddle area? (This might be suggestive of possible cauda equina compression.)

7. Does coughing or sneezing affect your symptoms? (Increased intra-abdominal pressure typically aggravates disc symptoms.)

8. Have you experienced any numbness or tingling in both hands or both feet at the same time? Any loss of coordination with gait? (This may suggest possible cord compression.)

☐ *History of the Current Episode*

It is essential to interrogate the patient very carefully regarding the history of the current problem. A thorough history may provide you with many clues about the nature of the problem as well as its progression. This will help you be most efficient when determining a treatment plan and setting goals. Ask the patient when and how this episode began (did it come on "all of a sudden" or was it gradual?). Also ask the patient to identify what symptoms appeared first and then how they progressed. Any treatment to date for this episode as well as its effects should be noted. Finally, ask the patient to assess the evolution of the problem, in other words, is it getting better, getting worse, or staying the same?

☐ *Previous History*

It is also important to note whether the patient has experienced any prior episodes of this nature and how they compare, in terms of both intensity and duration, to this event. Determine whether the location of the symptoms has varied across episodes. If there has been previous treatment for this problem, what was it and what were the effects, if any? Did the patient improve after the last episode and to what extent (100%, 50%, etc.)? If the patient reports less than 100% improvement, ask him or her to explain further so that the prognosis for this occurrence can be better determined. Any other significant medical history, including prior surgeries or hospitalizations, should be identified at this time.

PART

II

HEAD AND SPINE

SECTION 3

TEMPOROMANDIBULAR JOINT

This section addresses the temporomandibular joint or TMJ. This is a complex joint that is closely related to the upper cervical spine. Therefore it is essential that when examining a patient with this type of dysfunction the therapist rule out the cervical spine as a source of the patient's symptoms and/or as an associated factor.

JOINT BASICS

Articulation

The temporal bone along the surfaces of the articular eminence, glenoid fossa, and posterior glenoid spine articulates with the disc, which in turn articulates with the mandibular condyle (figure 3.1).

Type of Joint

The condyle of the mandible and inferior surface of the disc form a ginglymus joint; the superior surface of the disc and the articular eminence constitute an amphiarthrodial joint.

Degrees of Freedom

The axis of rotation through the lower joint is generally a line passing through both poles of the condyle (it is important to remember that anatomical variances may occur, even from one side to another in the same individual, that may affect this axis). Translation and gliding occur along the inferior surface of the temporal bone and the superior surface of the articulating disc.

Active Range of Motion

Mouth opening (mandibular depression): 40-55 mm (2-3 flexed proximal interphalangeal joints). To correctly calculate mouth opening, must add degree of *overbite* or vertical overlap of maxillary teeth to mandibular teeth (measure in millimeters).

Mouth closing (mandibular elevation): complete approximation of teeth.

Lateral deviation: approximately one-fourth opening range (generally 8-10 mm); symmetrical.

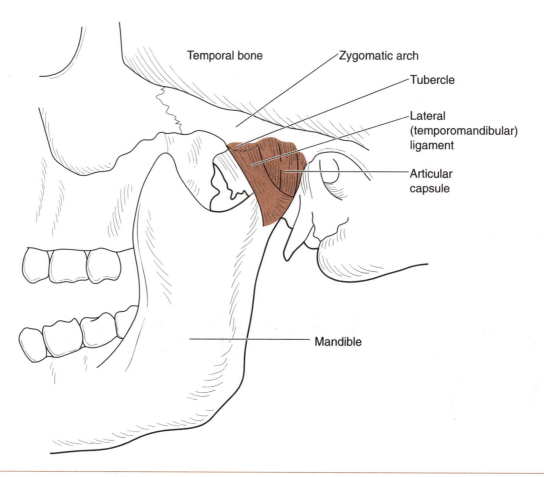

Figure 3.1 Temporomandibular joint.

Protrusion (mandible moves anteriorly): 3-6 mm. To correctly calculate protrusion, must add degree of *overjet* or anterior-posterior distance between overlapping maxillary and mandibular teeth (measure in millimeters).

Retrusion (mandible moves posteriorly): 3 mm

During active range of motion (AROM) assessment, must look for any deviation as well as limitation in range:

1. A C-curve during opening may indicate joint hypomobility on the side of deviation.
2. An S-curve during opening may suggest a muscle imbalance and/or capsulitis.

Accessory Movement

Distraction (Unilateral)

Assessment: To assess and/or increase joint mobility.

Patient: Sitting or supine with jaw relaxed.

Operator position: Place gloved thumb (opposite side being worked on) on patient's back molars with fingers outside mouth and encircled about jaw. Also support patient's head with other hand if patient is sitting.

Mobilizing force: Caudal direction; may be used for pain, mobility, disc reduction (if patient tolerates).

Distraction With Anterior Glide (Unilateral)

Assessment: To assess and/or increase joint mobility.

Patient: As above.

Operator position: As above.

Mobilizing force: As above, but after distracting, add anterior translation force; may be used for similar purposes.

Lateral Glide (Unilateral)

Assessment: To assess and/or increase joint mobility.

Patient: As above.

Operator position: As above.

Mobilizing force: Lateral force applied against molars with thumb; simultaneous medial force against body of mandible with fingers (via supination-type motion); may be used for similar purposes, especially medial disc displacement.

End-Feel

Mouth opening: *Firm* (tissue stretch secondary to tension on capsule, ligaments)

Mouth closing: Bone on bone (teeth approximate)

Protrusion: *Firm* (tissue stretch of posterior portion of disc)

Retrusion: *Firm* (tissue stretch of temporomandibular ligament)

Lateral deviation: *Firm*

Capsular Pattern

Limitation in mouth opening.

Close-Packed Position

Both extremes of range of motion.

1. Maximum opening: anterior joint close-packed limited by soft tissue and capsule
2. Maximum closing: posterior joint close-packed limited by teeth

Loose-Packed Position

Midrange at *freeway space.* Soft tissue of TMJ is the most relaxed, the mandible is slightly depressed, and the tongue rests against the hard palate (roughly 2-4 mm space between maxillary and mandibular teeth). The examiner may palpate by placing a finger (pad forward) in each external auditory canal with patient's mouth open and feeling for the point at which the mandibular condyle makes contact with the finger when patient closes mouth (figure 3.2, *a* and *b*).

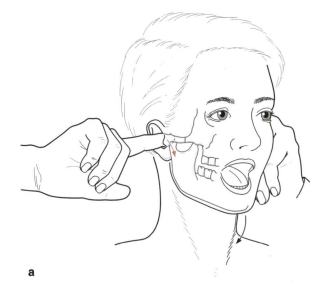

a

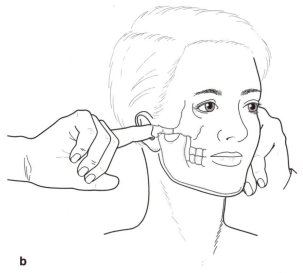

b

Figure 3.2 (*a*) To palpate the TMJ, place your index finger in the auditory canal. (*b*) At the freeway space, the examiner feels for the point at which the mandibular condyle makes contact with the finger on closing.

Stability

Capsular Ligament

Attaches at mandibular fossa, articular tubercle, and mandibular condyle; helps provide stability for disc atop mandible (figure 3.3).

Temporomandibular Ligament

Thickening of joint capsule that provides lateral reinforcement; attaches at zygomatic arch articular eminence and neck of mandible; restrains motion of lower jaw and prevents impingement of tissues behind mandibular condyle (figure 3.3).

Sphenomandibular Ligament

From the spine of the sphenoid to the middle ramus of the mandible; provides suspension for the mandible and prevents excessive anterior translation (figure 3.3).

Stylomandibular Ligament

A thickening of deep cerebral fascia running from the styloid process of the temporal bone to the posterior ramus of the mandible and inserting between the medial pterygoid and the masseter muscles; may assist in limiting protrusion (figure 3.3.).

The sphenomandibular and stylomandibular ligaments together are thought to keep the mandibular condyle, articular disc, and temporal bone opposed.

Disc

Biconcave shape provides increased congruency of joint in various positions; the thin center and wider anterior-posterior dimensions allow increased adaptability of the disc to its bony articular surfaces, thereby creating greater congruency.

Special Tests

Chvostek's Test

Assessment: Pathology of cranial nerve VII.

Patient: Sitting or supine.

Test position: Jaw relaxed, mouth closed.

Operator: Tap parotid gland over the masseter muscle.

Positive finding: Twitching of facial muscles.

Reciprocal Clicking

Assessment: Anterior displacement of disc.

Patient: Sitting or supine.

Test position: Begin with mouth closed; perform full opening followed by full closing.

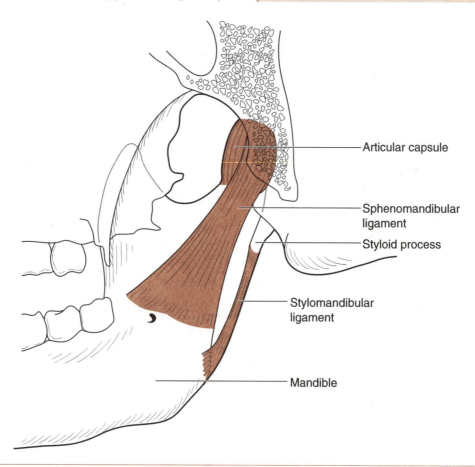

Articular capsule

Sphenomandibular ligament

Styloid process

Stylomandibular ligament

Mandible

Figure 3.3 TMJ ligaments.

> *Operator:* Palpate at lateral poles of condyles.
>
> *Positive finding:* Click is felt during opening and again during closing (may be at different positions); opening click generally louder of the two.

ARTHROKINEMATICS

Mandibular Depression

The convex condyles rotate anteriorly on a concave articular eminence during the first 10-25 mm of mandibular opening. After 10-25 mm, the condyle-disc complex translates anteriorly in conjunction with continued rotation to permit full mouth opening.

Mandibular Elevation

During mouth closing, the reverse actions take place. The condyle-disc complex translates posteriorly followed by posterior rotation of the condyle on the disc.

Protrusion

Both mandibular condyles along with the discs translate anteriorly to produce this motion.

Retrusion

Both mandibular condyles along with the discs translate posteriorly to produce this motion.

Lateral Deviation

The condyle on the ipsilateral side spins while the condyle on the contralateral side translates anteriorly.

Distraction

Distraction occurs ipsilaterally at condyle during biting unilaterally against resistance (such as food) placed between upper and lower third molars.

Compression

Compression occurs contralaterally at condyle during biting unilaterally against resistance (such as food) placed between upper and lower third molars. Compression may also occur secondary to muscular contraction during functional activities.

Lateral glide is an accessory motion that may accompany any or all of the arthrokinematics described to provide more available movement at the joint.

NEUROLOGY

The neurology of the TMJ region is rather complex; this explains why it is easy to confuse pain of TMJ origin with other types of facial pain. Therapists should have a general idea of the motor and sensory nerves that supply the general region.

Motor

The majority of the muscles around the TMJ are supplied by the mandibular branch of the fifth cranial nerve (temporalis, medial and lateral pterygoid, and masseter) and by the seventh cranial nerve (digastric, stylohyoid).

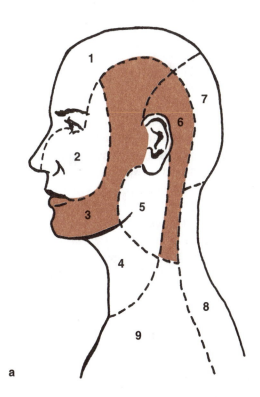

Figure 3.4 (*a*) Sensory nerve distribution of head, neck, and face: (1) Ophthalmic nerve, (2) maxillary nerve, (3) mandibular nerve, (4) transverse cutaneous nerve of neck (C2-C3), (5) greater auricular nerve (C2-C3), (6) lesser auricular nerve (C2), (7) greater occipital nerve (C2-C3), (8) cervical dorsal rami (C3-C5), and (9) suprascapular nerve (C5-C6).

(continued)

SECTION **3** TMJ

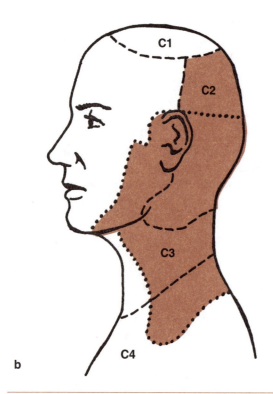

Figure 3.4 (*b*) Dermatome pattern of the head, neck, and face. C3 is shown in dotted lines because of overlap.

Sensory

The auriculotemporal and masseteric branches of the mandibular nerve supply the majority of sensory innervation to the TMJ.

Structures innervated by the afferents of cervical vertebrae 1 through 3 can all refer pain to the head, TMJ, and face. Additionally, the ganglia of cranial nerve V is located at about C3, thus presenting the possibility for cervical pathology to result in pain referrals along the sensory distribution of the trigeminal nerve.

SURFACE PALPATION

TMJ (via external auditory canal)

Mandibular condyles

Pterygoid, temporalis, and masseter muscles

Mandible

Teeth

Hyoid bone (anterior to C2-C3)

Thyroid cartilage (anterior to C4-C5)

Mastoid process

Cervical spine (for details see guide to cervical spine, section 4)

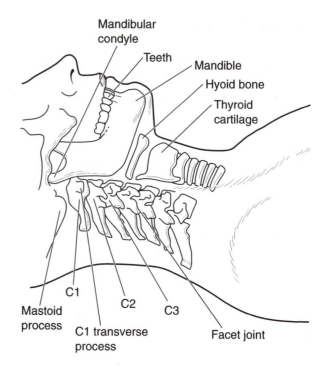

Figure 3.5 Anatomy of the jaw and neck.

MUSCLE ORIGIN AND INSERTION

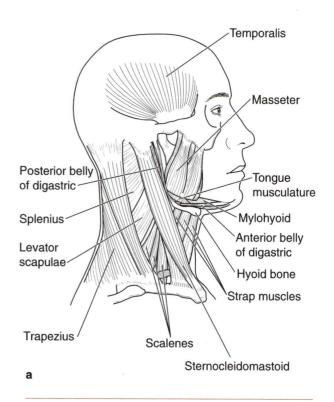

Figure 3.6 Muscles of the jaw and neck.

Table 3.1   Muscle Origin and Insertion

Muscle	Origin and insertion
Lateral pterygoid	Pterygoid process and base of skull *to* mandibular neck and articular disk
Mylohyoid	Mylohyoid line of body of mandible *to* hyoid bone
Geniohyoid	Inferior mental spine of mandible *to* hyoid bone
Digastric	Mastoid process of temporal bone *to* posterior chin (body of mandible) with slip of fascia connecting tendon to hyoid bone
Masseter	Zygomatic arch *to* ramus and angle of mandible
Temporalis	Lateral skull *to* coronoid process and anterior ramus of mandible
Medial pterygoid	Pterygoid process *to* angle of mandible
Stylohyoid	Styloid process of temporal bone *to* hyoid bone

SECTION
3

TMJ

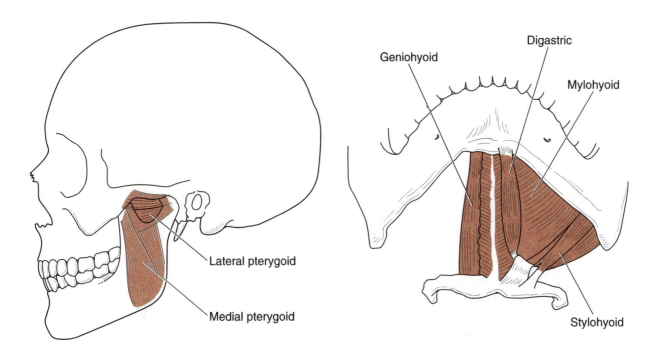

b

c

Figure 3.6   *(continued)*

MUSCLE ACTION AND INNERVATION

Table 3.2 Muscle Action and Innervation

Action	Muscle involved	Nerve supply
Mandibular depression (mouth opening)	Lateral pterygoid Mylohyoid* Geniohyoid* Digastric*	Mandibular (CN V) Inferior alveolar (CN V) Hypoglossal (CN XII) Inferior alveolar (CN V) Facial (CN VII)
Mandibular elevation (mouth closing)	Masseter Temporalis Medial ptergyoid	Mandibular (CN V) Mandibular (CN V) Mandibular (CN V)
Mandibular protrusion	Lateral pterygoid Medial pterygoid Masseter* Mylohyoid* Geniohyoid* Digastric* Stylohyoid* Temporalis (anterior fibers)*	Mandibular (CN V) Mandibular (CN V) Mandibular (CN V) Inferior alveolar (CN V) Hypoglossal (CN XII) Inferior alveolar (CN V) Facial (CN VII) Facial (CN VII) Mandibular (CN V)
Mandibular retraction	Temporalis (posterior fibers) Masseter* Digastric* Stylohyoid* Mylohyoid* Geniohyoid*	Mandibular (CN V) Mandibular (CN V) Inferior alveolar (CN V) Facial (CN VII) Facial (CN VII) Inferior alveolar (CN V) Hypoglossal (CN XII)
Mandibular deviation (lateral)	Lateral pterygoid (ipsilateral) Medial pterygoid (contralateral) Temporalis* Masseter*	Mandibular (CN V) Mandibular (CN V) Mandibular (CN V) Mandibular (CN V)

* Act only when assistance is required. CN = cranial nerve.

Table 3.3 Differential Diagnosis

Disorder	Description	Onset	Symptoms	Signs	Special tests
Synovitis	Inflammation of synovium; may be accompanied by joint effusion	Gradual or acute trauma	Pain in preauricular area; increased or decreased with functional activities; better with rest	May be unable to fully close or open secondary to pain; pain with TMJ palpation while closing	(+) For pain contralateral side with TMJ dynamic loading (forced biting); (+) pain with passive loading (retrusive overpressure)

Table 3.3 *(continued)*

Disorder	Description	Onset	Symptoms	Signs	Special tests
Capsulitis	Inflammation of joint capsule	Gradual or acute trauma	Pain in preauricular area; increased or decreased with functional activities; better with rest	May be unable to fully open secondary to pain; pain with TMJ palpation with mouth closed and with mouth opened 30 mm	(+) For pain ipsilateral side with TMJ distraction (forced biting); (+) for pain with passive loading (retrusive overpressure)
Fibrosis	Fibrosis of joint capsule	Follows history of prolonged capsulitis, prolonged immobilization or mandibular restriction, trauma or repetitive microtrauma, arthritis	Stiffness possibly accompanied by pain	Decreased mandibular mobility and deflection from midline suggesting decrease in arthrokinematic translation	
Hyper-mobility	Excessive joint mobility	May report history of "catching" when mouth opened making closing difficult	Feeling of jaw "going out of place," may describe joint noises	Large indentation palpable posterior to condyle when mouth opened, deviation of mandible to contralateral side at end opening; depression > 40 mm	
Dislocation	Joint locked in "open" position	May report history of "catching" or feeling of jaw going out of place; may or may not have pain	Inability to close mouth	Mouth opened and deviated to contralateral side	
Disc displacement with reduction	Disc rests in dislocated position, is reduced with mouth opening and returns to position with mouth closing	Trauma or gradual onset	Reports Joint noises during mouth opening and closing; may describe a click with opening and again with closing	Plapation over lateral poles reveals clicks at different points opening closing; opening click generally louder	Therapist lifts angle of mandible anterior/superior while patient opens and close; checking to see if "noise" and/or pain

(continued)

Table 3.3 *(continued)*

Disorder	Description	Onset	Symptoms	Signs	Special tests
					enchanced, which supports diagnosis of displacement with reduction
Disc displacement without reduction (acute)	Displacement of disc anteriorly that does not reduce with mandibular motion	Previous history of joint noise and intermittent locking	Inability to open mouth fully; difficulty with functional activities such as yawning, chewing	Mandibular depression and protrusion limited with deviation toward the involved side; lateral devia-tion limited toward the opposite side	
Disc displacement without reduction (chronic)	Same as above	Same as above	Not functionally limited but complains of joint noise with motion (crepitus)	As above; however, limit-ations in range are very slight if present at all; palpable crepitus is present	
Osteo-arthritis	Degenerative joint disease	Gradual following prolonged inflammatory condition or after acute trauma	Same as for inflammatory conditions (capsulitis, synovitis); also with joint noise	Same as for inflammatory conditions (capsulitis, synovitis) *but* palpable crepitus	(+) X-ray for structural changes
Osteo-arthrosis	Degenerative joint disease— further progression	End result of osteoarthritis	Primary complaint is noise; pain often decreases	Close to functional depression and protrusion with slight deviation toward involved side at end ranges; close to functional lateral deviation to opposite side	(+) X-ray for structural changes
Fibrous ankylosis	Fibrous adhesions in joint	Usually a result of major trauma and/or surgery	Multidirectional limitation in movement	Multidirectional limitation in movement— generally a result of decreased translation	

CERVICAL SPINE

The cervical spine (figure 4.1, *a* and *b*) may be responsible for symptoms in the neck, shoulder, arm, head, or face. It is important for therapists to have an appreciation of the many possible presentations of cervical spine dysfunction.

JOINT BASICS

Articulation

There are two unique articulations within the cervical spine: the occipitoatlantal (OA) joint between the skull and C1 and the atlantoaxial (AA) joint between C1 and C2. At the OA, the two convex occipital condyles articulate with the two concave superior facets of the atlas. The AA joint is actually made up of three joints: the median AA (atlanto-odontoid articulation) and two lateral joints. The median AA is the articulation of the dens and the atlas, and the lateral joints are the articulations between the convex inferior facets of the atlas and the concave superior facets of the axis. Vertebrae C3-C7 are intervertebral joints much like the remaining joints in the spine. The articulations occur between the intervertebral body and the intervertebral disc, as well as between the right and left superior facets of one vertebra with the right and left inferior facets of the body above. Cervical vertebrae are unique in that on the anterior aspect bilaterally they possess an additional articulation. The uncovertebral joints, or joints of von Luschka, are not present at birth but develop around the age of 10 as a result of upright weight bearing. The right and left superior uncovertebral joints of one cervical vertebra articulate with the right and left inferior uncovertebral joints of the cervical body above (figure 4.2).

Facet Orientation

AA joint: The facets lie parallel to the transverse plane, thus permitting rotation around a vertical axis.

C3-C7 facets: The superior facets are oriented upward, posteriorly and medially, while the inferior facets are oriented downward, anteriorly and laterally.

Type of Joint

The OA joint is a plane synovial joint. At the AA joint, the median AA is a synovial trochoid joint whereas the lateral joints are plane synovial joints. For the remaining spinal segments, a cartilaginous joint exists between the vertebral body and the disc, while the articulation of the superior articular process (facet) and the inferior articular process (facet) constitutes a plane diarthrodial joint.

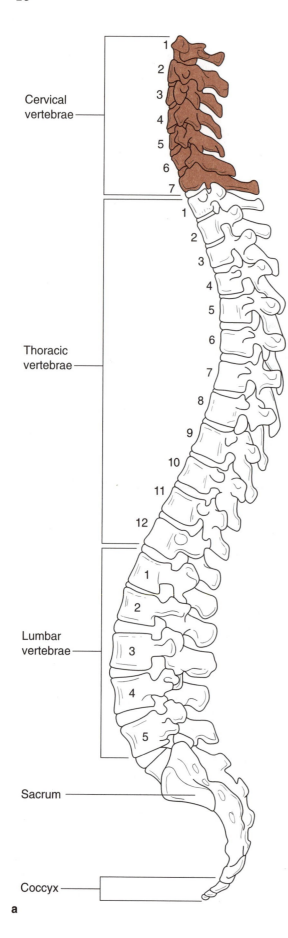

Cervical vertebrae

Thoracic vertebrae

Lumbar vertebrae

Sacrum

Coccyx

a

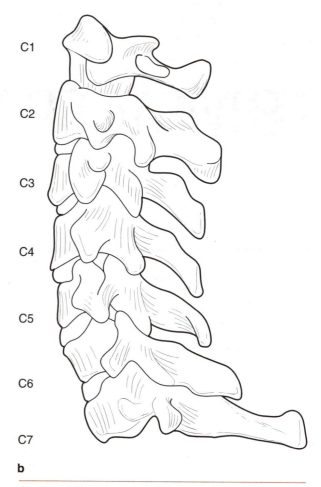

C1

C2

C3

C4

C5

C6

C7

b

Figure 4.1 (*a*) Spinal column: cervical vertebrae. (*b*) Cervical vertebrae (lateral view).

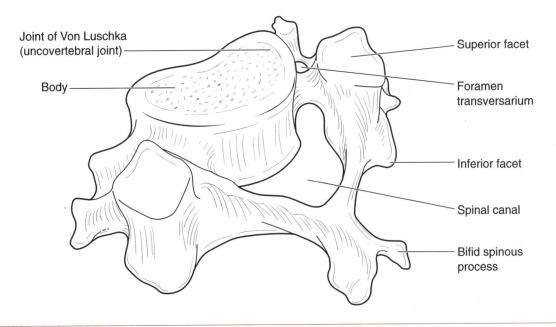

Joint of Von Luschka
(uncovertebral joint)

Body

Superior facet

Foramen
transversarium

Inferior facet

Spinal canal

Bifid spinous
process

Figure 4.2 Typical cervical vertebra.

Degrees of Freedom

- Flexion and extension in a sagittal plane about a coronal axis.
- Side-bending in a frontal plane about a sagittal axis.
- Rotation in a transverse plane about an axis in the frontal and sagittal planes.

The motions described may occur at any one motion segment; however, because vertebral motion segments function as part of an articular system, motion between any two vertebrae is generally limited to a small amount of translation (or glide) and rotation. Furthermore, because each intervertebral joint functions as part of an articular system, the axis of motion of any given joint changes in response to motion at segments above and below.

Active Range of Motion

OA: approximately 17° of flexion, 12° of extension, 3-5° of side-bending, 3-5° of rotation (2-3° with initial motion, 2-3° more at end of physiological rotation range)

AA: 47° rotation, 10° flexion and extension, little to no side-bending

C3-C7: 45° rotation, 40° flexion, 24° extension, 50° side-bending (fairly evenly distributed among these segments)

Gross cervical motion: 80-90° flexion, 70° extension, 20-45° sidebending, 70-90° rotation

Accessory Movement

Central Posteroanterior Pressure (Central P-A)

Assessment: To assess and/or increase central P-A mobility.

Patient: Prone.

Operator position: At patient's head, place thumb pads over spinous process with hands supporting soft tissue.

Mobilizing force: Posterior to anterior.

Unilateral Posteroanterior Pressure (Unilateral P-A)

Assessment: To assess and/or increase unilateral P-A mobility.

Patient: Prone.

Operator position: At patient's head, place thumb pads over facet to be mobilized with hands supporting soft tissue.

Mobilizing force: Posterior to anterior with slight angulation to replicate facet orientation.

Transverse Pressure

Assessment: To assess and/or increase rotation.

Patient: Prone.

Operator position: At side to be mobilized, place thumbs adjacent to spinous process.

Mobilizing force: Transverse (perpendicular to spinous process) away from examiner's body.

Manual Traction

Assessment: To unload spinal segments.

Patient: Supine (alternate position may be used if patient cannot tolerate supine).

Operator position: At patient's head with hands on posterior neck at segment level where traction is desired.

Mobilizing force: Gentle distraction force (cephalad).

End-Feel

Flexion: *Firm*
Extension: *Firm*
Side-bending: *Firm*
Rotation: *Firm*

Capsular Pattern

Side-bending and rotation limited equally, then extension.

Close-Packed Position

Extension (facets).

Loose-Packed Position

Slightly extended.

Stability

Transverse Atlantal Ligament (Atlantal Cruciform Ligament)

Traverses the ring of the atlas dividing it into a larger posterior section that houses the spinal cord and a smaller anterior space for the dens. Longitudinal fibers extend superiorly to attach to the occiput; inferior fibers descend to the posterior portion of the axis; the transverse portion maintains the dens and the atlas in close approximation. The primary function of this ligament is to prevent anterior displacement of C1 on C2 (figure 4.3).

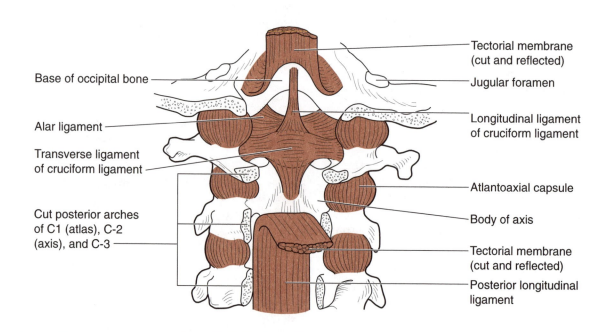

Figure 4.3 Cervical ligaments.

Alar Ligaments

Travel from either side of the dens to the medial aspect of the occiput; they are taut in flexion but relaxed in extension. The left ligament checks rotation of the head and neck to the right; the right lower portion and the left upper portion limit side-bending of the head and neck to the right. This ligament also functions to prevent distraction of C1 on C2.

Posterior AA

This is a continuation of the ligamentum flavum in the upper cervical; runs from posterior arch of atlas to arch of axis.

Anterior AA

This is a continuation of the anterior longitudinal ligament in the upper cervical; runs from anterior-inferior atlas to anterior axis.

Tectorial Membrane

This is the continuation of the posterior longitudinal ligament above the axis; thin membrane between occiput and C2.

Ligamentum Nuchae

This ligament runs posteriorly from C1 and connects all cervical spinous processes.

Special Tests

Vertebral Artery

Assessment: Compromise of the vertebral artery (either stretch or compression).

Patient: Sitting or supine with head supported in therapist's hands.

Test position: Rotate head to end of available range; then bring head into extension; patient's eyes remain open throughout; at end range have patient slowly count to 15.

Operator: Support patient's head and observe any changes.

Positive finding: Any combination of the following: nystagmus, vertigo, changes in voice or halting of voice, pupillary dilation and constriction, personality changes; if positive findings occur, discontinue physical therapy until further evaluation by physician.

Transverse Ligament Stability Test

Assessment: Integrity of transverse ligament.

Patient: Supine on table.

Test position: Anterior glide of C1 on C2.

Operator: One hand supports head; the other hand grasps around mass of C1 with thumb and index finger; glide anteriorly.

Positive finding: No firm end-feel; patient may have symptoms of dizziness, sensation of choking, numbness in face.

Passive Neck Flexion

Assessment: Dural tension.

Patient: Supine, no pillow if tolerable.

Test position: Passively flex patient's head until encountering resistance or symptoms.

Operator: Stand to patient's side, supporting head.

Positive finding: Resistance encountered before end of range, reproduction of patient's symptoms; note that mild pulling or stretching sensation in lower cervical or midthoracic is a normal response.

ARTHROKINEMATICS

Upper Cervical

Flexion

Occiput glides posteriorly on atlas; atlas then glides forward and tilts anteriorly.

Extension

Occiput glides anteriorly on atlas; atlas glides backward, then tilts posteriorly.

Side-Bending

Occipital condyles glide opposite direction of side-bending; atlas then slides ipsilaterally and rotates contralaterally secondary to tension in alar ligament.

SECTION 4 Cervical Spine

Rotation

Occiput rotates only 2-3° on atlas; then occiput and atlas rotate together to same side; tension on alar ligament causes occiput to side-bend contralaterally near end range.

Mid and Lower Cervical

The cervical facets (C2-C7) are oriented at a 45° angle to the transverse plane.

Flexion

The disc is compressed anteriorly; facets glide cranially.

Extension

The disc is compressed posteriorly; facets glide caudally.

Side-Bending

The disc is compressed on the side of the concavity; facets on the side of the concavity glide caudally while those contralateral to concavity glide cranially.

Rotation

The disc undergoes torsion; facets glide in the same direction as vertebral body rotation.

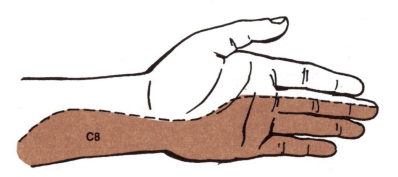

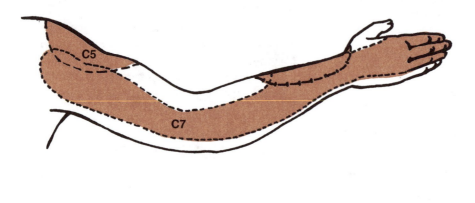

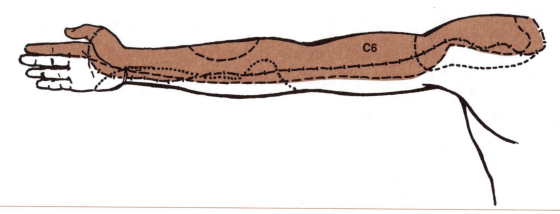

Figure 4.4 Dermatomes of C4 through T1.

NEUROLOGY

Table 4.1 Neurology

Nerve root	Reflex	Motor	Sensory
C1		Rectus capitis anterior and lateral	
C2		Rectus capitis posterior	Side of head
C3		Scalenes, erector spinae	Anterior and lateral neck
C4		Levator scapulae, trapezius	Lateral neck to shoulder
C5	Biceps	Deltoid	Lateral shoulder
C6	Brachioradialis	Biceps, extensor carpi Radialis longus and brevis	Posterior and radial aspect of thumb
C7	Triceps	Triceps, flexor carpi radialis	Long finger
C8	Abductor digiti minimi	Flexor digitorum profundus, extensor pollicis longus	Ulnar aspect of little finger
T1		Hand intrinsics	Medial forearm and arm

SURFACE PALPATION

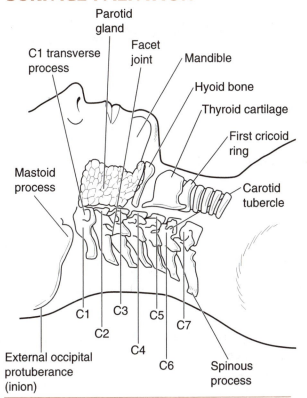

Posterior
External occipital protuberance
Inion
Cervical spinous processes
Cervical facets
Mastoid process

Lateral
Cervical transverse processes
Temporomandibular joint
Mandible
Parotid

Anterior
Hyoid
Thyroid
First cricoid ring
First rib
Supraclavicular fossa

Figure 4.5 Anatomy of neck and cervical spine.

MUSCLE ORIGIN AND INSERTION

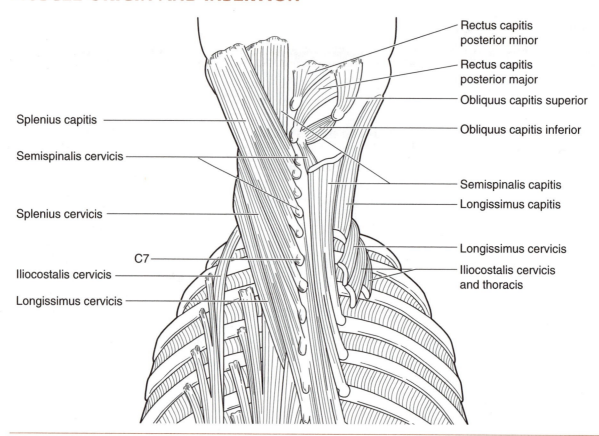

Figure 4.6 Major muscles of the neck.

Table 4.2 Muscle Origin and Insertion

Muscle	Origin and insertion
Rectus capitis anterior	Occiput *to* anterior portion of lateral mass of C1
Rectus capitis lateralis	Occiput *to* transverse process of C1
Longus capitis	Inferior occiput *to* transverse processes of C3-C6
Obliquus capitis superior	Transverse process of C1-T4 *to* lateral portion of inferior nuchal line
Sternocleidomastoid	Clavicle and sternum *to* mastoid process
Splenius capitis nuchal	Spinous processes C2-T4 *to* lateral portion of superior line and mastoid process
Semispinalis capitis	Transverse processes C7-T6 and articular processes of C4-C6 *to* superior and inferior nuchal lines
Longissimus capitis	Transverse processes T1-T5 and articular processes of C4-C7 *to* mastoid process

Table 4.2 *(continued)*

Muscle	Origin and insertion
Spinalis capitis	Medial portion of semispinalis capitis (see previous page)
Trapezius	Occiput, ligamentum nuchae, and spinous processes of C7-T12 *to* lateral 1/3 of the clavicle, scapular spine, and acromion
Rectus capitis posterior major	Spinous process of C2 *to* inferior nuchal line
Rectus capitis posterior minor	Posterior arch of C1 *to* inferior nuchal line
Obliquus capitis inferior	Spinous process of C2 *to* transverse process of C1
Longus colli	Anterior vertebral bodies T3-midcervical *to* transverse processes of vertebrae above OR anterior bodies of vertebrae above up to anterior arch atlas
Scalenus anterior	Transverse processes of C3-C6 *to* upper surface of 1st rib (near sternum)
Scalenus medius	Transverse processes of C2-C7 *to* 1st rib
Scalenus posterior	Transverse processes of C5-C6 *to* 2nd rib
Splenius cervicis	Spinous processes of C2-T4 *to* transverse processes of C1-C3
Semispinalis cervicis	Transverse processes of T1-T6 *to* spinous processes C2-C5
Longissimus cervicis	Transverse processes of T1-T5 *to* transverse processes of C2-C6
Levator scapulae	Superior angle of scapula *to* transverse processes of C2-C4
Iliocostalis cervicis	Upper 6 ribs *to* transverse processes of C4-C6
Rotatores	Transverse processes *to* spinous processes from sacrum to C2; each fascicle spans only 1-2 segments
Multifidus	Transverse processes *to* spinous processes from sacrum to C2; each fascicle spans only 2-4 segments
Intertransversarii	Transverse processes *to* spinous processes from sacrum C2; each fascicle spans only 1-2 segments

SECTION

4

Cervical Spine

MUSCLE ACTION AND INNERVATION

Table 4.3   Muscle Action and Innervation

Action	Muscle involved	Nerve supply
Upper cervical flexion	Rectus capitis anterior	C1-C2
	Rectus capitis lateralis	C1-C2
	Longus capitis	C1-C3
	Hyoid muscles	Inferior alveolar nerve
		Facial nerve
		Hypoglossal nerve
		Ansa cervicalis
	Obliquus capitis superior	C1
	Sternocleidomastoid (if head neutral or flexed)	Accessory C2
Upper cervical extension	Splenius capitis	C4-C6
	Semispinalis capitis	C1-C8
	Longissimus capitis	C6-C8
	Spinalis capitis	C6-C8
	Trapezius	Accessory C3-C4
	Rectus capitis posterior minor	C1
	Rectus capitis posterior major	C1
	Obliquus capitis superior	C1
	Obliquus capitis inferior	C1
	Sternocleidomastoid (if head slightly extended)	Accessory C2
Upper cervical rotation (muscles contract unilaterally)	Trapezius (face moves to opposite side)	Accessory C3, C4
	Splenius capitis (face moves to same side)	C4-C6
	Longissimus capitis (face moves to same side)	C6-C8
	Semispinalis capitis (face moves to same side)	C1-C8
	Obliquus capitis inferior (face moves to same side)	C1
	Sternocleidomastoid (face moves to opposite side)	Accessory C2
Upper cervical side-bending (muscles contract unilaterally)	Trapezius	Accessory C3, C4
	Splenius capitis	C4-C6
	Longissimus capitis	C6-C8
	Semispinalis capitis	C1-C8
	Obliquus capitis inferior	C1
	Rectus capitis lateralis	C1-C2
	Longus capitis	C1-C3
	Sternocleidomastoid	Accessory C2
Neck flexion	Longus colli	C2-C6
	Scalenus anterior	C4-C6
	Scalenus medius	C3-C8
	Scalenus posterior	C6-C8

Table 4.3 *(continued)*

Action	Muscle involved	Nerve supply
Neck extension	Splenius cervicis	C6-C8
	Semispinalis cervicis	C1-C8
	Longissimus cervicis	C6-C8
	Levator scapulae	C3-C4 Dorsal scapular
	Iliocostalis cervicis	C6-C8
	Spinalis cervicis	C6-C8
	Multifidus	C1-C8
	Interspinalis cervicis	C1-C8
	Trapezius	Accessory C3-C4
	Rectus capitis posterior major	C1
	Rotatores brevis	C1-C8
	Rotatores longi	C1-C8
Side-bending of neck	Levator scapulae	C3-C4 Dorsal scapular
	Splenius cervicis	C6-C8
	Iliocostalis cervicis	C6-C8
	Longissimus cervicis	C6-C8
	Semispinalis cervicis	C1-C8
	Multifidus	C1-C8
	Intertransversarii	C1-C8
	Scaleni	C3-C8
	Sternocleidomastoid	Accessory C2
	Obliquus capitis inferior	C1
	Rotatores brevis	C1-C8
	Rotatores longi	C1-C8
	Longus colli	C2-C6
Neck rotation (muscles contract unilaterally)	Levator scapulae (face moves to same side)	C3-C4 Dorsal scapular
	Splenius cervicis (face moves to same side)	C6-C8
	Iliocostalis cervicis (face moves to same side)	C6-C8
	Longissimus cervicis (face moves to same side)	C6-C8
	Semispinalis cervicis (face moves to same side)	C1-C8
	Multifidus (face moves to opposite side)	C1-C8
	Intertransversarii (face moves to same side)	C1-C8
	Scaleni (face moves to opposite side)	C3-C8
	Sternocleidomastoid (face moves to opposite side)	Accessory C2
	Obliquues capitis inferior (face moves to same side)	C1
	Rotatores brevis (face moves to same side)	C1-C8
	Rotatores longi (face moves to same side)	C1-C8

SECTION **4**

Cervical Spine

DIFFERENTIAL DIAGNOSIS

Essentially with the spine, it is not feasible to offer complete guidelines for differential diagnosis. Any pathology in the upper extremity or in the lower extremity could be of spinal origin or could have some relationship to spinal dysfunction; the practitioner always needs to keep this in mind.

SECTION 5

THORACIC SPINE

The thoracic spine is a unique region because of the rib articulations with each thoracic vertebra (figure 5.1). The ribs connect the thoracic spine to the sternum, and the cage that is formed houses multiple vital organs including the lungs and heart. Because of the many visceral structures that are in proximity to the thoracic region, it is important for the clinician to recognize visceral pain referral patterns to assist with differential diagnoses. Likewise, many musculoskeletal complaints of thoracic origin may mimic visceral symptoms, so the clinician needs to be able to perform a complete assessment of this region.

JOINT BASICS

Articulation

There are two primary types of articulations along the thoracic spine itself: one between the vertebral body and the intervertebral disc and another between the right and left superior facets and the right and left inferior facets of the body above (figure 5.2). Other joints in the thoracic region (considering ribs, costal cartilage, manubrium, and sternum) include the following.

Costotransverse

Convex costal tubercle of rib articulates with concave costal facet on transverse processes of T1-T10; there are no costotransverse joints at T11-T12 because ribs 11 and 12 do not articulate with transverse processes.

Costovertebral

Convex head of rib articulates with two concave demifacets on adjacent thoracic vertebrae; ribs 2-10 fit in this angle created by two demifacets and have some contact with the intervertebral disc; ribs 1, 11, and 12 articulate with one vertebra only (figure 5.3).

Costochondral

Ribs 1-7 (true ribs) articulate with costal cartilage.

Chondrosternal

The cartilage articulating with ribs 1-7 in turn articulates with the manubriosternum.

Interchondral

The costal cartilage of ribs 8-10 (false ribs) articulates with the cartilage above them (that of ribs 1-7), thereby attaching them to the sternum via a fused costal cartilage.

Manubriosternal

The manubrium articulates with the sternum.

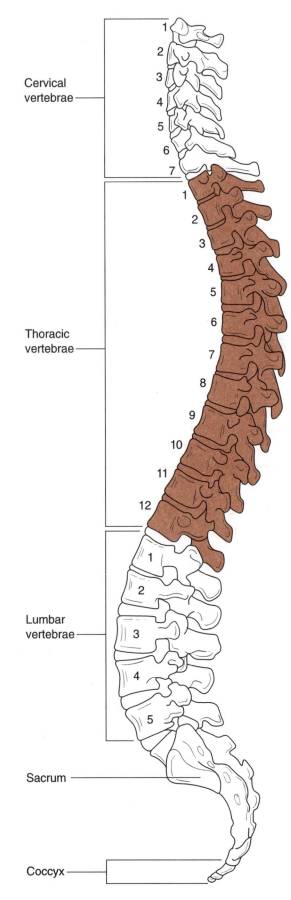

a

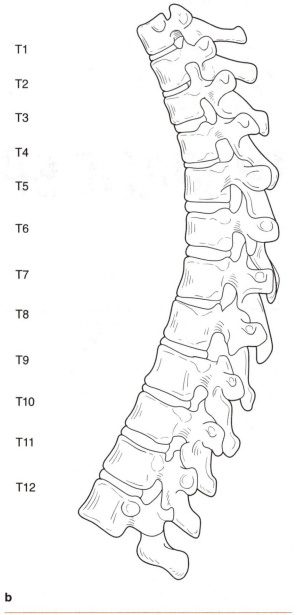

b

Figure 5.1 (*a*) Spinal column: thoracic vertebrae. (*b*) Thoracic vertebrae (lateral view).

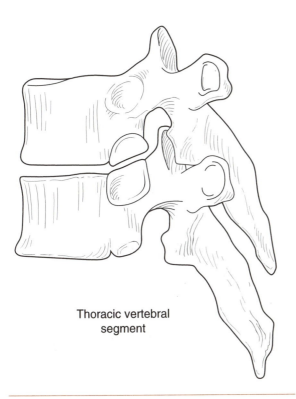

Figure 5.2 Thoracic vertebral segment.

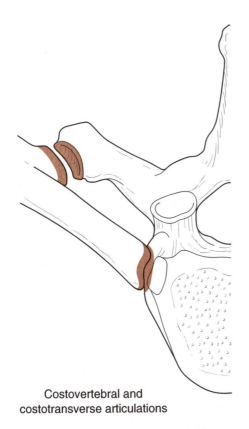

Costovertebral and
costotransverse articulations

Figure 5.3 Costovertebral and costotransverse articulations (shaded areas).

Xiphisternal

The body of the sternum articulates with the xiphoid process.

Type of Joint

A thoracic spinal segment constitutes a symphysis between the vertebral body and the disc and a plane diarthrodial joint between the superior articular process (facet) of one vertebra and the inferior articular process (facet) of the vertebra above.

> Costotransverse: plane synovial
>
> Costovertebral: plane synovial
>
> Costochondral: synchondrosis
>
> Chondrosternal: synovial
>
> Interchondral: synovial
>
> Manubriosternal: synchondrosis
>
> Xiphisternal: synchondrosis

Degrees of Freedom

- Flexion and extension in a sagittal plane about a coronal axis.
- Side-bending in a frontal plane about a sagittal axis.
- Rotation in a transverse plane about an axis in the frontal and sagittal planes.

The motions described may occur at any one motion segment; however, because vertebral motion segments function as part of an articular system, motion between any two vertebrae is generally limited to a small amount of translation (or glide) and rotation. Furthermore, because each intervertebral joint functions as part of an articular system, the axis of motion of any given joint changes in response to motion at segments above and below.

Active Range of Motion

The active range of motion includes gross thoracic motion; individual spinal segmental motion cannot be measured except by x-ray.

> Flexion: 20-45°
>
> Extension: 25-45°
>
> Side-bending: 20-40°
>
> Rotation: 35-50°

Costovertebral (Rib Cage) Expansion: 3 to 7.5 cm

Costovertebral expansion is measured by placing a tape measure around the individual's chest at the level of the fourth intercostal space. The patient begins by exhaling as much as possible, at which point the initial measurement is made. The patient then inhales maximally and a second measurement is made. The difference between the two is then noted.

Rib Motion

Rib motion is not quantifiable numerically. An examiner should assess it by placing hands over patient's upper chest and feeling for anteroposterior rib motion

- Inspiration: Ribs 1 through 6 elevate and increase anteroposterior diameter (pump-handle motion); ribs 7 through 10 elevate and move laterally to increase lateral dimension (bucket-handle motion).
- Expiration: Is the reverse of inspiration.

Flexion

Assessment: To assess segmental forward bending in thoracic spine (figure 5.4)

Patient: Sitting with arms across chest.

Operator position: Stand at patient's side with one hand supporting underneath patient's arms and the other hand palpating the interspinous space at the desired level; palpate gapping as patient's trunk flexes passively.

Mobilizing force: None.

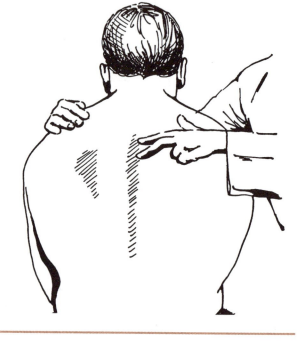

Figure 5.4 Mobility testing—thoracic flexion.

Extension

Assessment: To assess segmental backward bending in thoracic spine (figure 5.5).

Patient: Sitting with arms crossed in front and hands on opposite shoulders, elbows forward.

Operator position: Stand at patient's side with one hand supporting underneath patient arms and the other hand palpating the interspinous space at the desired level; palpate approximation as patient's trunk extends passively.

Mobilizing force: None.

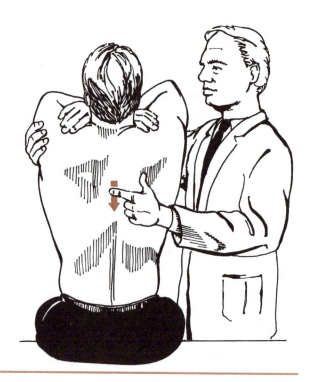

Figure 5.5 Mobility testing—thoracic extension.

Side-Bending

Assessment: To assess segmental side-bending in thoracic spine (figure 5.6).

Patient: Sitting with arms across chest.

Operator position: Stand at patient's side with one hand supporting across arms (patient's shoulder in therapist axilla) and the other hand palpating the interspinous space at the desired level; palpate gapping or approximation as patient's trunk is passively side-bent.

Mobilizing force: None.

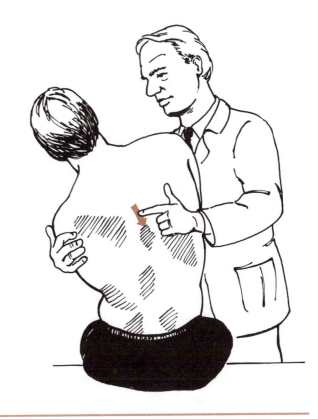

Figure 5.6 Mobility testing—thoracic side-bending.

Rotation

Assessment: To assess segmental rotation in thoracic spine (figure 5.7).

Patient: Sitting with arms across chest.

Operator position: Stand at patient's side with one hand supporting across arms (patient's shoulder in therapist axilla) and the other hand palpating the interspinous space at the desired level; palpate motion of spinous process above relative to spinous process below.

Mobilizing force: None.

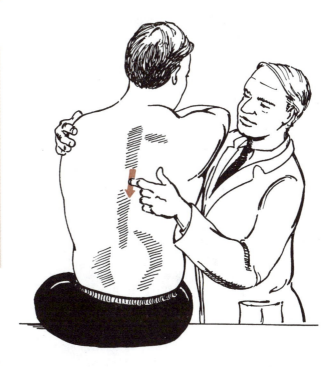

Figure 5.7 Mobility testing—thoracic rotation.

Accessory Movement

Central Posteroanterior Pressure (Central P-A)

Assessment: To assess and/or increase central P-A mobility.

Patient: Prone.

Operator position: At patient's head or side, place thumb pads on spinous process with hands supporting soft tissues.

Mobilizing force: Posterior-to-anterior pressure.

Rib Spring

Assessment: To assess rib mobility.

Patient: Prone.

Operator position: At patient's side (facing patient's side, head, or feet depending on rib angulation at segment being evaluated) with hands around rib cage (approximately 45° to vertical).

Mobilizing force: Direct pressure anteriorly as well as slightly inward to compress rib and then release.

Traction

Assessment: To unload spinal segments.

Patient: Sitting with hands clasped behind head and elbows forward for upper thoracic or arms crossed across body for lower thoracic.

Operator position: Stand behind patient with knees slightly flexed; place arms between patient's arms and forearms, grasping onto the forearms for upper thoracic and around patient's arms (above shoulders and grasping forearms) for lower thoracic.

Mobilizing force: Straighten knees and lean slightly back to provide traction force.

Transverse Pressure

Assessment: To assess and/or increase rotation.

Patient: Prone.

Operator position: At side to be mobilized, place thumb pads adjacent to spinous process.

Mobilizing force: Transverse pressure (perpendicular to spinous processes) directed away from examiner's body.

Unilateral Posteroanterior Pressure (Unilateral P-A)

Assessment: To assess and/or increase unilateral P-A mobility.

Patient: Prone.

Operator position: At patient's head or side to be mobilized, place thumb pads pointing toward each other over transverse process.

Mobilizing force: Posterior-to-anterior pressure.

End-Feel

Flexion: *Firm* (tension in posterior longitudinal ligament, ligamentum flavum, interspinous ligaments, and facet joint capsules)

Extension: Bone on bone; *firm* due to contact of spinous processes and facets; disc size; tension in anterior longitudinal ligament, joint capsule, and abdominal muscles

Side-bending: Bone on bone (limited by facets and rib cage)

Rotation: Bone on bone (limited structurally by rib cage)

Capsular Pattern

Side-bending and rotation limited equally, then extension.

Close-Packed Position

Extension (facets).

Loose-Packed Position

Midway between flexion and extension.

Stability

The ligamentum flavum and the anterior longitudinal ligament are thicker in the thoracic region, thereby enhancing stability here. Facet joint capsule laxity is decreased in the thoracic spine. There is generally decreased flexibility and increased stability in the thoracic region secondary to structural features such as the presence of ribs, elongated spinous processes, tighter facet capsules, thicker ligamentum flavum, and increased vertebral body dimensions.

Special Tests

Slump

Assessment: Neural tension.

Patient: Sitting on edge of table.

Test position: Guide the patient through the following steps one at a time, stopping if encountering resistance or reproduction of patient's symptoms:

- Patient sits with knees together, brings hands behind back to rest on table.
- Patient rounds shoulders (sacrum should remain vertical).
- Patient flexes head and neck; if no complaints, therapist applies gentle pressure through upper back and shoulders to help patient maintain position.
- Therapist uses one hand to passively extend one of patient's legs.
- Therapist passively dorsiflexes foot.

Operator: Stand at patient's side, performing as above.

Positive finding: Early resistance or reproduction of patient's symptoms at any point throughout the test; confirm by releasing pressure at shoulders while keeping leg extended; if patient elevates head and symptoms decrease and leg extends farther (if not already at full extension and/or dorsiflexion), then positive for neural tension.

SECTION 5

Thoracic Spine

Passive Neck Flexion

Assessment: Dural tension.

Patient: Supine, no pillow if tolerable.

Test position: Passively flex patient's head until encountering resistance or symptoms.

Operator: Stand to patient's side, supporting head.

Positive finding: Resistance encountered before end of range, reproduction of patient's symptoms; note that mild pulling or stretching sensation in lower cervical and/or midthoracic is a normal response.

Upper Limb Tension Tests
(See All Tests in Section 7)

Thoracic Outlet Tests

Assessment: Compression of the neurovascular bundle at the thoracic outlet; often characterized by compression of any or all of the following: subclavian artery, subclavian vein, brachial plexus.

Positive finding: Various tests are performed in an effort to target different structures; however, the tests alone should not be used to diagnose but rather to support clinical presentation. During all tests the patient is observed for the following signs: reproduction of symptoms, decreased radial pulse, trophic changes, pallor, cyanosis.

Adson (Scalene) Maneuver

Patient: Sitting.

Test position: Patient inhales deeply and holds breath, extends neck and rotates head toward side being examined.

Operator: Maintain upper extremity in slight abduction and palpate radial pulse.

Hyperabduction Maneuver

Patient: Sitting.

Test position: Patient's arm raised overhead in abduction.

Operator: Maintain upper extremity in abduction and palpate radial pulse.

Costoclavicular Maneuver

Patient: Sitting.

Test position: Patient retracts shoulders (military position).

Operator: Palpate radial pulse.

ARTHROKINEMATICS

Flexion

The disc is compressed anteriorly; facets glide cranially.

Extension

The disc is compressed posteriorly; facets glide caudally.

Side-Bending

The disc is compressed on the side of the concavity; facets on the side of the concavity glide caudally while those contralateral to concavity glide cranially.

Rotation

The disc undergoes torsion; facets glide relative to one another; direction depends on relative position of spine (i.e., neutral, flexed, extended).

NEUROLOGY

There are 12 thoracic nerves on each side of the body (see figure 5.8). The first 11 lie between the ribs, and the 12th lies below rib 12. These 12 nerves, or intercostal nerves, supply primarily the thoracic and abdominal walls; however, the first two thoracic nerves also provide some innervation to the upper limb. Note that the sympathetic ganglia in the thoracic region lie in close proximity to the heads of the ribs.

SURFACE PALPATION

Anterior Chest

Sternum
Manubrium
Body
Xiphoid
Clavicle
Abdominal quadrants
Ribs (anterior and lateral)
Sternocostal cartilages
Costochondral cartilages

Posterior Chest

Ribs (posterior and lateral)
Scapulae
Spine (spinous processes, facets, muscles, ligaments)

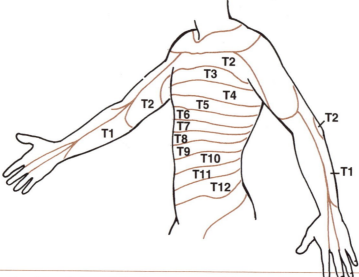

Figure 5.8 An anterolateral view of the dermatomes of the spine.

MUSCLE ORIGIN AND INSERTION

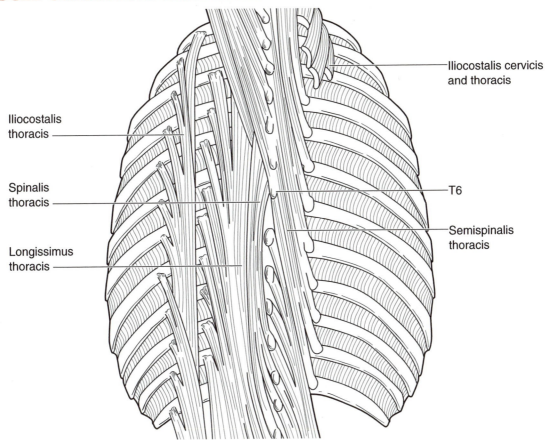

Iliocostalis cervicis
and thoracis

Iliocostalis
thoracis

Spinalis
thoracis

Longissimus
thoracis

T6

Semispinalis
thoracis

Figure 5.9 Major muscles of the midback.

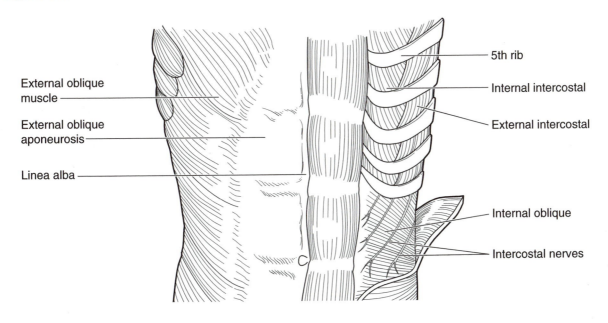

External oblique
muscle

External oblique
aponeurosis

Linea alba

5th rib

Internal intercostal

External intercostal

Internal oblique

Intercostal nerves

Figure 5.10 Musculature of the anterolateral abdominal wall.

Table 5.1 Muscle Origin and Insertion

Muscle	Origin and insertion
Rectus abdominis	Pubis *to* sternum and lower costal cartilages
External abdominal oblique	Lower 6 ribs *to* anterior iliac crest, pubis, and linea alba
Internal abdominal oblique	Iliopsoas fascia, anterior iliac crest, and lumbar fascia *to* lower 3 ribs, xiphoid process, linea alba, and symphysis pubis
Spinalis thoracis	Common tendon of erector spinae and lower thoracic spinous processes *to* upper thoracic spinous processes
Iliocostalis thoracis	Upper borders of lower 6 ribs *to* lower borders of upper 6 ribs
Longissimus thoracis	Common tendon of erector spinae *to* lower 9-10 ribs and adjacent transverse processes
Semispinalis thoracis	Transverse processes of thoracic vertebrae *to* spinous processes of upper thoracic and lower cervical vertebrae
Multifidus	Transverse processes *to* spinous process from sacrum to C2; each fascicle spans only 2-4 segments
Rotatores	Transverse processes *to* spinous processes from sacrum to C2; each fascicle spans only 1-2 segments
Interspinalis	Shorter group of rotatores *to* spinous processes of contiguous vertebrae
Intertransversii	Shorter group of rotatores *to* transverse processes of contiguous vertebrae
Transversus abdominis	Lower 6 ribs, lumbar fascia, iliac crest, and inguinal ligament *to* xiphoid process, linea alba, symphysis pubis
Scalenus anterior	Transverse process of C3-C6 *to* upper surface of 1st rib (near sternum)
Scalenus medius	Transverse processes of C2-C7 *to* 1st rib
Scalenus posterior	Transverse processes of C5-C6 *to* 2nd rib
Serratus posterior superior	Lower ligamentum nuchae, spinous process of C7-T3 *to* ribs 2 to 5
Iliocostalis cervicis	Upper 6 ribs *to* transverse processes of C4-C6
Levatores costarum	Transverse processes of C7-T11 *to* upper edge of rib below
Pectoralis major	Medial 2/3 of clavicle, sternum, upper 6 ribs *to* lateral intertubercular groove

Table 5.1 *(continued)*

Muscle	Origin and insertion
Pectoralis minor	Ribs 3-5 *to* coracoid process of scapula
Serratus anterior	Anterolateral ribs 1-8 *to* costal surface of medial border of scapula
Sternocleidomastoid	Clavicle and sternum *to* mastoid process
Serratus posterior inferior	Lower 2 thoracic and upper 2 lumbar spinous processes *to* lower 4 ribs
Iliocostalis lumborum	Iliac crest and sacrum *to* lower 6-7 ribs
Quadratus lumborum	Medial iliac crest and last rib *to* lumbar transverse processes
Intercostals	Rib *to* rib all levels
Diaphragm	Inner surfaces of sternum and lower ribs and upper lumbar bodies *to* central tendon that divides thoracic and abdominal cavities
Latissimus dorsi	Lower 6 thoracic and all lumbar and sacral spinous processes and iliac crest *to* medial lip of intertubercular groove

SECTION 5

Thoracic Spine

MUSCLE ACTION AND INNERVATION

Table 5.2 Muscle Action and Innervation

Action	Muscle involved	Nerve supply
Thoracic flexion	Rectus abdominis External abdominal oblique (bilateral) Internal abdominal oblique (bilateral)	T6-T12 T7-T12 T7-T12, L1
Thoracic extension	Spinalis thoracis Iliocostalis thoracis (bilateral) Longissimus thoracis (bilateral) Semispinalis thoracis (bilateral) Multifidus (bilateral) Rotatores (bilateral) Interspinalis	T1-T12 T1-T12 T1-T12 T1-T12 T1-T12 T1-T12 T1-T12
Thoracic rotation and side-bending	Iliocostalis thoracis (same side) Longissimus thoracis (same side) Intertransversii (same side) Internal abdominal oblique (same side) Semispinalis thoracis (opposite side) Multifidus (opposite side) Rotatores (opposite side)	T1-T12 T1-T12 T1-T12 T7-T12, L1 T1-T12 T1-T12 T1-T12

(continued)

Table 5.2 *(continued)*

Action	Muscle involved	Nerve supply
	External abdominal oblique (opposite side)	T7-T12
	Transversus abdominis (opposite side)	T7-T12, L1
Rib elevation	Scalenus anterior (1st rib)	C4-C6
	Scalenus medius (1st rib)	C3-C8
	Scalenus posterior (2nd rib)	C6-C8
	Serratus posterior superior (ribs 2-5)	Intercostal 2-5
	Iliocostalis cervicis (ribs 1-6)	C6-C8
	Levatores costarum (all ribs)	T1-T12
	Pectoralis major (arm fixed)	Later pectoral (C6-C7) Medial pectoral (C7-C8, T1)
	Serratus anterior (lower ribs if scapula fixed)	Long thoracic (C5-C7)
	Pectoralis minor (ribs 2-5 if scapula fixed)	Lateral pectoral (C6-C7) Medial pectoral C7-C8, T1)
	Sternocleidomastoid (if head fixed)	Accessory (C3-C3)
Rib depression	Serraus posterior inferior lower 4 ribs)	T9-T12
	Iliocostalis lumborum (lower 6 ribs)	L1-L3
	Longissimus thoracis	T1-T12
	Rectus abdominis	T6-T12
	External abdominal oblique (lower 5-6 ribs)	T7-T12
	Internal abdominal oblique (lower 5-6 ribs)	T7-T12, L1
	Transversus abdominis	T7-T12, L1
	Quadratus lumborum (12th rib)	T12, L1-L4
	Transversus thoracis	T1-T12
Rib approximation	Iliocostalis thoracis	T1-T12
	Intercostals (internal and external)	Intercostal 1-11
	Diaphragm	Phrenic
	External intercostals	Intercostal 1-11
	Transverse thoracis (sternocostals)	Intercostal 1-11
	Diaphragm	Phrenic
	Sternocleidomastoid	Accessory (C2-C3)
	Scalenus anterior	C4-C6
	Scalenus medius	C3-C8
	Scalenus posterior	C6-C8
	Pectoralis major	Lateral pectoral (C6-C7) Medial pectoral (C7-C8, T1)
	Pectoralis minor	Lateral pectoral (C6-C7) Medial pectoral (C7-C8, T1)

Table 5.2   *(continued)*

Action	Muscle involved	Nerve supply
	Serratus anterior	Long thoracic (C5-C7)
	Latissimus dorsi	Thoracodorsal (C6-C8)
	Serratus posterior superior	Intercostal 2-5
	Iliocostalis thoracis	T1-T12
Expiration	Internal intercostals	Intercostal 1-11
	Rectus abdominis	T6-T12
	External abdominal oblique	T7-T12
	Internal abdominal oblique	T7-T12, L1
	Iliocostalis lumborum	L1-L3
	Longissimus	T1-L3
	Serratus posterior inferior	T9-T12
	Quadratus lumborum	T12, L1-L4

SECTION
5

Thoracic Spine

SECTION
6

LUMBAR SPINE

This section addresses the lumbar spine (figure 6.1, *a* and *b*). Low back pain is one of the most common complaints of patients who present to physical therapy. Therefore, the ability of the therapist to perform a thorough assessment of the area and to differentiate the origin of various spinal pathologies is essential. The therapist should also be suspicious of lumbar involvement whenever a patient complains of lower-extremity pain, tingling, or weakness of unknown etiology. The spine is often overlooked in these cases; however, when correctly identified as the problem source, it may be very responsive to physical therapy interventions.

JOINT BASICS

Articulation

There are two primary types of articulations along the lumbar spine: the one between the vertebral body and the intervertebral disc and the other between the right and left superior facets and the right and left inferior facets of the body above. There are five pairs of facets in the lumbar region. The facets from L1 to L4 are oriented primarily in the sagittal plane, with the concave superior facets facing medially and posteriorly and the convex inferior facets directed anteriorly and laterally. The facets at L5 shift their orientation primarily to the frontal plane, and they are more widely spaced than those of the lumbar vertebrae above.

Type of Joint

A lumbar spinal segment constitutes a cartilaginous joint between the vertebral body and the disc and a plane diarthrodial joint between the superior articular process (facet) of one vertebra and the inferior articular process (facet) of the vertebra above.

Degrees of Freedom

- Flexion and extension in a sagittal plane about a coronal axis.
- Side-bending in a frontal plane about a sagittal axis.
- Rotation in a transverse plane about an axis in the frontal and sagittal planes.

These motions may occur at any one motion segment; however, because vertebral motion segments function as part of an articular system, motion between any two vertebrae is generally limited to a small amount of translation (or glide) and rotation. Furthermore, because each intervertebral joint functions as part of an articular system, the axis of motion of any given joint changes in response to motion at segments above and below.

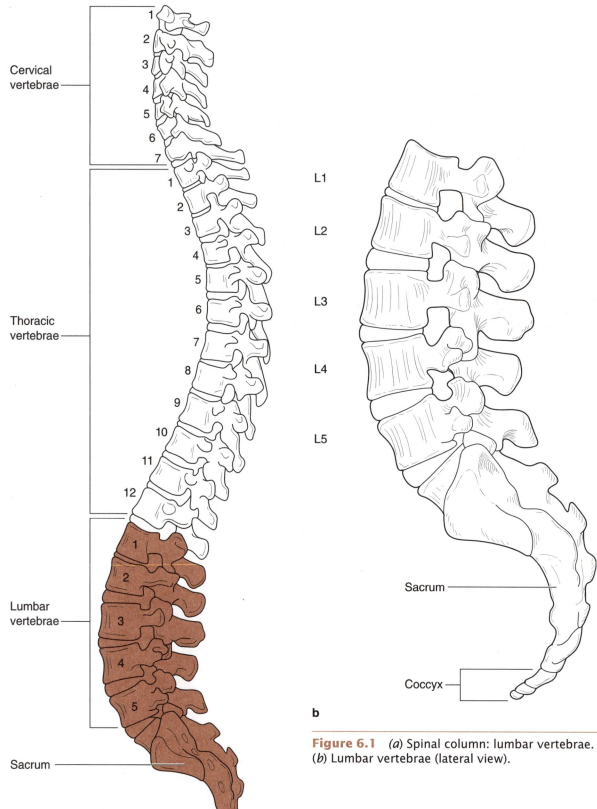

Cervical
vertebrae

1
2
3
4
5
6
7

Thoracic
vertebrae

1
2
3
4
5
6
7
8
9
10
11
12

Lumbar
vertebrae

1
2
3
4
5

Sacrum

Coccyx

a

L1

L2

L3

L4

L5

Sacrum

Coccyx

b

Figure 6.1 (*a*) Spinal column: lumbar vertebrae.
(*b*) Lumbar vertebrae (lateral view).

Active Range of Motion

The L1-L4 facet orientation is such that rotation and side-bending are more limited and flexion and extension are more available. Of the latter two, flexion is typically more limited in the lumbar region. The exception is at the L5-S1 articulation, which has little to no rotation but constitutes the majority of the flexion in the lumbar region. The following list describes gross lumbar motion; individual spinal segmental motion cannot be measured except by x-ray.

Flexion: 40-60°

Assessment: To assess segmental forward bending in lumbar spine.

Patient: Side-lying.

Operator position: Stand facing patient with one hand supporting underneath patient leg and the other hand palpating the interspinous space at the desired level; palpate gapping as patient's trunk is passively flexed.

Extension: 20-35°

Assessment: To assess segmental backward bending in lumbar spine.

Patient: Side-lying.

Operator position: Stand facing patient with one hand supporting underneath patient leg and the other hand palpating the inter-spinous space at the desired level; palpate approximation as patient's trunk is passively extended.

Rotation: 3-18°

Assessment: To assess segmental rotation in lumbar spine.

Patient: Prone.

Operator position: Stand at patient's side with one hand at patient's opposite iliac crest and the other hand palpating the interspinous space at the desired level; palpate motion of spinous process above relative to spinous process below.

Side-Bending: 15-20°

Assessment: To assess segmental side-bending in lumbar spine.

Patient: Prone.

Operator position: Stand at patient's side with one hand supporting patient's leg and the other hand palpating the interspinous space at the desired level; palpate gapping or approximation as patient's trunk is passively side-bent by abducting leg.

OR

Patient: Side-lying.

Operator position: Stand facing patient with one hand supporting patient's legs (at the ankles) and the other hand palpating the interspinous space at the desired level; palpate gapping or approximation as patient's trunk is passively side-bent by raising or lowering both legs.

Combined Movements

Extension and rotation (quadrant)
Side-bending in flexion
Side-bending in extension
Flexion and rotation

Accessory Movement

Central Posteroanterior Pressure (Central P-A)

Assessment: To assess and/or increase central P-A mobility.

Patient: Prone.

Operator position: At patient's side, place thumb pads or distal pisiform contact over spinous process.

Mobilizing force: Posterior to anterior.

Unilateral Posteroanterior Pressure (Unilateral P-A)

Assessment: To assess and/or increase unilateral P-A mobility.

(continued)

Patient: Prone.

Operator position: At patient's side to be mobilized; thumb pads pointing toward each other over transverse process on the same side.

Mobilizing force: Posterior to anterior.

Transverse Pressure

Assessment: To assess and/or increase rotation.

Patient: Prone.

Operator position: At patient's side to be mobilized; thumb pads adjacent to spinous process on the same side.

Mobilizing force: Transverse (perpendicular to spinous processes) away from examiner's body.

Traction

Assessment: To unload spinal segments.

Patient: Supine hook-lying.

Operator position: Kneel on table at patient's feet with towel or sheet behind patient's knees.

Mobilizing force: Lean backward to provide traction force.

End-Feel

Flexion: *Firm* (tension in posterior longitudinal ligament, ligamentum flavum, interspinous ligaments, and facet joint capsules)

Extension: Bone on bone, *firm* (contact of spinous processes, facets, disc size, tension in anterior longitudinal ligament, joint capsule, and abdominal muscles)

Side-bending: *Firm* (facet approximation, rib cage, joint capsule tension)

Rotation: *Firm* (impact of articular processes, intertransverse ligaments)

Capsular Pattern

Side-bending and rotation limited equally (may be to same or opposite sides, depending on segmental level), then extension.

Close-Packed Position

Extension (facets).

Loose-Packed Position

Midway between flexion and extension.

Stability

Together the presence of the following ligaments in the lumbar spine helps contribute to overall stability:

1. Anterior longitudinal ligament attaches anteriorly from vertebral body above to vertebral body below.

2. Posterior longitudinal ligament attaches posteriorly from bony rim to bony rim with some attachments to the disc as well.

3. Ligamentum flavum lamina to lamina with attachments to facet capsule.

4. Interspinous runs between spinous processes.

5. Supraspinous runs superiorly along spinous processes.

6. Iliolumbar ligament runs from transverse processes of L4 and L5 to iliac crests; composed of five bands—anterior, posterior, superior, inferior, and vertical; broad, strong ligament that helps to stabilize the fifth lumbar vertebra on the sacrum (prevent anterior displacement). This ligament is present only in adults and is thought to develop from the lower fibers of the quadratus lumborum as a result of age-related changes and stress associated with upright posture and ambulation.

7. Thoracolumbar fascia surrounds the muscles of the lumbar spine; composed of three layers—anterior, middle, and posterior. The anterior layer or *passive* layer is a thin layer thought to be derived from the fascia of the quadratus lumborum. The middle layer lies posterior to the quadratus lumborum. The posterior or *active* layer is a fascial layer that fuses with the transversus abdominis, thereby providing an indirect attachment for this muscle to the lumbar vertebrae via the spinous processes. Tension in the fascia is thus transmitted to the spinous processes of L1-L4 and may assist the spinal extensor musculature in resisting an applied load.

The annulus, facet joint capsules, facet structural configurations, and increased disc and vertebral body size in the lumbar region all help to contribute to stability.

Special Tests

Passive Neck Flexion

Assessment: Dural tension.

Patient: Supine, no pillow if tolerable.

Test position: Passively flex patient's head until encountering resistance or symptoms.

Operator: Stand to patient's side, supporting head.

Positive finding: Resistance encountered before end of range, reproduction of patient's symptoms; note that mild pulling or stretching sensation in lower cervical and mid-thoracic is a normal response.

Straight Leg Raise

Assessment: Neural tension, hamstring length.

Patient: Supine, arms at side, preferably without pillow if tolerable.

Test position: Gradually lift patient's leg into hip flexion keeping knee extended; may use *sensitization* to enhance tension along various nerve pathways by adding one or several of the following components: passive neck flexion, dorsiflexion and plantarflexion, inversion and eversion, internal and external rotation, and adduction.

Operator: Stand close to patient's side while performing the above; observe patient for indications of discomfort while also assessing any changes in resistance through motion.

Positive finding: Reproduction of patient's back or leg symptoms for tension, tightness in muscle belly for tight hamstring. With any sensitization tests, the therapist is looking for reproduction of symptoms along specific nerve pathway.

Slump

Assessment: Neural tension.

Patient: Sitting on edge of table.

Test position: Guide the patient through the following steps one at a time, stopping if encountering resistance or reproduction of patient's symptoms:

- Patient sits with knees together, brings hands behind back to rest on table.
- Patient rounds shoulders (note: sacrum should remain vertical).
- Patient flexes head and neck; if no complaints, therapist applies gentle pressure through upper back and shoulders to help patient maintain position.
- Therapist uses one hand to passively extend one of patient's legs.
- Therapist passively dorsiflexes foot.

Operator: Stand at patient's side, performing as above.

Positive finding: Early resistance or reproduction of patient's symptoms at any point throughout the test; confirm by releasing pressure at shoulders while keeping leg extended; if patient elevates head and symptoms decrease and leg extends farther (if not already at full extension and/or dorsiflexion), then positive for neural tension.

Prone Knee Bend (Femoral Nerve Stretch)

Assessment: Femoral nerve tension.

Patient: Prone.

Test position: Hip in extension (as patient lies flat on the table).

Operator: Passively flex knee; can increase tension by further hip extension.

Positive finding: Reproduction of anterior leg symptoms or low back pain; suggests possible involvement of femoral nerve and/or lumbar nerve roots 2-4.

Faber

Assessment: Range of motion of hip and pain.

Patient: Supine with leg placed so foot is resting on top of knee of opposite leg.

Test position: Hip in position of flexion, abduction, external rotation.

Operator: Slowly lower test leg in abduction toward table.

Positive finding: Leg does not reach table or pain is present in this position.

Babinski

Assessment: Upper motor neuron lesion.

Patient: Supine and relaxed.

Test position: As above.

Operator: Use thumb or handle of reflex hammer; apply swift stroke along plantar surface of foot, moving from heel along lateral surface and then crossing the ball of the foot.

Positive finding: Great toe extension, possibly with splaying of other toes into abduction.

Sacroiliac Joint (SIJ) Clearing (Gap and Compression)

Assessment: Possible involvement of the SIJ.

Patient: Supine.

Test position: Operator's hands at top of iliac crests bilaterally.

Operator: To test gapping of the SIJ, squeeze iliac crests toward each other; to test compression of the SIJ, apply pressure directing iliac crests away from one another.

Positive finding: Reproduction of patient symptoms, particularly in location of SIJ.

Shift Assessment

Assessment: Possible disc or root involvement; patient tends to shift away from the painful side as a protective mechanism.

Patient: Standing; may be further assessed in supine position.

Test position: Stance.

Operator: View position of shoulders in relation to pelvis.

Positive finding: If shoulders are shifted to one side of the pelvis, then shift is present; shift is named by the side that the shoulders are leaning toward.

Shift Correction

Assessment: Provides information as to whether patient will tolerate attempts at correcting shift.

Patient: Standing; may also be done in supine or prone position, depending on patient comfort.

Test position: Stance, supine, prone.

Operator: Stance—place one hand at patient's shoulders on side patient is shifted toward, other hand on patient's pelvis on opposite side; gently attempt to bring shoulders and pelvis in line with one another. Supine or prone—place hands at patient's hips; gently lift patient's hips and slide in an attempt to bring hips in line with shoulders.

Positive finding: If patient can tolerate the position, then shift-correcting activities can and should be encouraged; if patient cannot tolerate the position (either markedly increased back pain or reproduction or increase of peripheral symptoms), then shift-correcting activities should not be promoted at this time.

ARTHROKINEMATICS

Flexion

The disc is compressed anteriorly; facets glide cranially.

Extension

The disc is compressed posteriorly; facets glide caudally.

Side-Bending

The disc is compressed on the side of the concavity; facets on the side of the concavity glide cau-dally while those contralateral to concavity glide cranially.

Rotation

The disc undergoes torsion; facets glide relative to one another; direction depends on the relative position of the spine (neutral, flexed, extended).

SECTION 6

Lumbar Spine

NEUROLOGY

Table 6.1 Neurology

Nerve root	Reflex	Motor	Sensory
L1			Inguinal region
L2		Hip flexors	Proximal anterior thigh
L3		Quadriceps	Middle anteromedial and thigh
L4	Patellar tendon	Tibialis anterior	Anteromedial thigh and knee
L5	Medial hamstring	Extensor hallucis longus, extensor digitorum	Lateral thigh, leg, dorsum of foot
S1	Achilles	Peroneals and hamstrings	lateral foot
S2		Toe flexors	Sole of foot

SURFACE PALPATION

Posterior

Lumbar spinous process
Lumbar facets, transverse processes
Sacral spine
Iliac crest
Posterior superior iliac spine
Coccyx
Ischial tuberosity
Sciatic nerve
Piriformis

Supraspinous ligament
Paraspinal muscles
Inferior lateral angles
Sacral sulcus

Anterior

Abdominal quadrants
Inguinal region
Iliac crest
Anterior superior iliac spine
Symphysis pubis
Greater trochanter

MUSCLE ORIGIN AND INSERTION

Table 6.2 Muscle Origin and Insertion

Muscle	Origin and insertion
Psoas major	Anterolateral aspect of lumbar vertebral bodies and transverse processes *to* below lesser trochanter
Rectus abdominis	Pubis *to* sternum and lower costal cartilages
External abdominal oblique	Lower 6 ribs *to* anterior iliac crest, pubis, and linea alba
Internal abdominal oblique	Iliopsoas fascia, anterior iliac crest, and lumbar fascia *to* lower 3 ribs, xiphoid process, linea alba, and symphysis pubis
Transversus abdominis	Lower 6 ribs, lumbar fascia, iliac crest, and inguinal ligament *to* xiphoid process, linea alba, symphysis pubis
Latissimus dorsi	Lower 6 thoracic and all lumbar and sacral spinous processes and iliac crest *to* medial lip of intertubercular groove
Erector spinae	Posterior surface of sacrum, iliac crest, and spinous processes of lumbar and last 2 thoracic vertebrae *to* all vertebrae and skull via 3 divisions
Transversospinalis	Rotator muscles from sacrum to C2; run transverse process *to* spinous process varying from 1-2 segments to 2-4 segments
Interspinales	Transverse processes *to* spinous processes from sacrum to C2; each fascicle spans only 1-2 segments
Quadratus lumborum	Medial iliac crest and last rib *to* lumbar transverse processes
Intertransversarii	Transverse processes *to* spinous processes from sacrum to C2; each fascicle spans only 1-2 segments

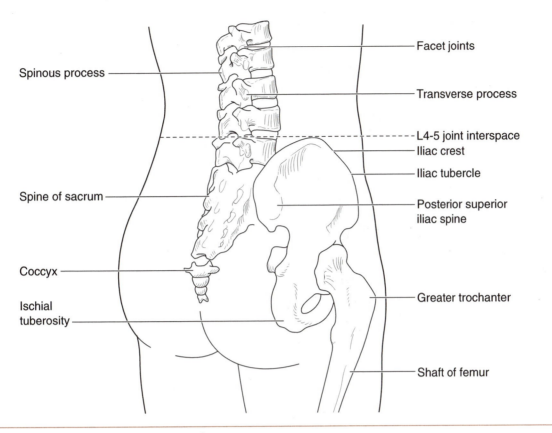

Figure 6.2 Posterior lumbar spine palpation.

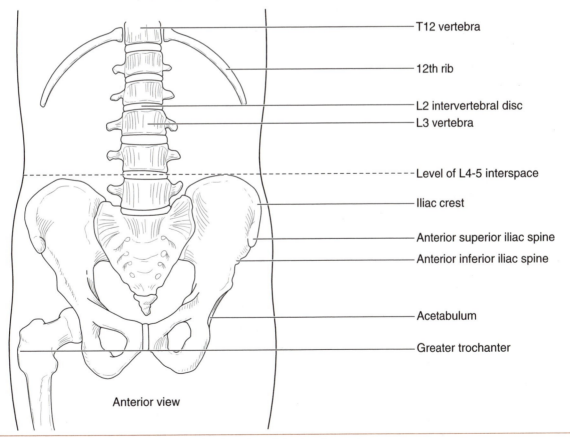

Figure 6.3 Anterior view of the boney landmarks of the lumbar spine.

(Figure 6.2 labels)
Facet joints
Spinous process
Transverse process
L4-5 joint interspace
Iliac crest
Iliac tubercle
Spine of sacrum
Posterior superior iliac spine
Coccyx
Greater trochanter
Ischial tuberosity
Shaft of femur

(Figure 6.3 labels)
T12 vertebra
12th rib
L2 intervertebral disc
L3 vertebra
Level of L4-5 interspace
Iliac crest
Anterior superior iliac spine
Anterior inferior iliac spine
Acetabulum
Greater trochanter
Anterior view

MUSCLE ACTION AND INNERVATION

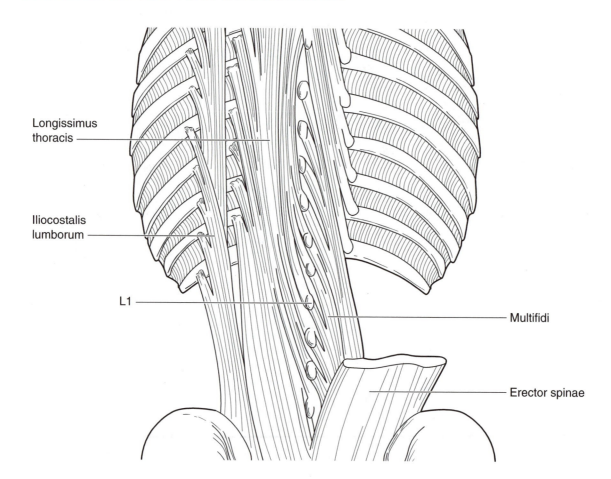

Longissimus
thoracis

Iliocostalis
lumborum

L1

Multifidi

Erector spinae

a

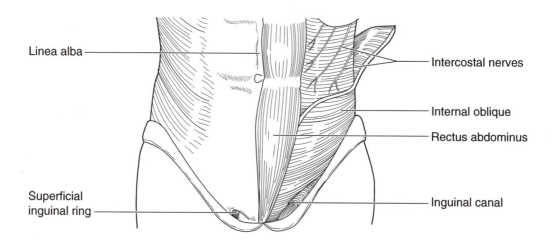

Linea alba

Intercostal nerves

Internal oblique

Rectus abdominus

Superficial
inguinal ring

Inguinal canal

b

Figure 6.4 (*a*) Major muscles of the lower back. (*b*) Musculature of the anterolateral abdominal wall.

Table 6.3 Muscle Action and Innervation

Action	Muscle involved	Nerve supply
Forward bending	Psoas major Rectus abdominis External abdominal oblique Internal abdominal oblique Transversus abdominis	L1-L3 T6-T12 T7-T12 T7-T12, L1 T7-T12, L1
Backward bending	Latissimus dorsi Erector spinae Transversospinalis Interspinales Quadratus lumborum	Thoracodorsal (C6-C8) L1-L3 L1-L5 L1-L5 T12, L1-L4
Side-bending	Latissimus dorsi Erector spinae Transversospinalis Interspinales Quadratus lumborum Psoas major External abdominal oblique	C6-C8 L1-L3 L1-L5 L1-L5 T12, L1-L4 L1-L3 T7-T12

DIFFERENTIAL DIAGNOSIS

Table 6.4 lists common musculoskeletal disorders of the spine seen by therapists. The clinical findings listed below each disorder represent possible findings rather than absolute findings. Use this table as a guideline to form a hypothesis of the nature of the problem. The patient may exhibit only partial components of a disorder or may be affected by a combination of disorders. This listing does not include all possible disorders including several pathological disorders that must be considered. Abbreviated treatment suggestions are included.

Table 6.4a Differential Diagnosis

Nature	Discogenic	Spinal stenosis	Instability (spondylolisthesis)	Strain or sprain
Area	Central Buttock area	Extremity symptoms (Sx) often > central Can be bilateral Asymmetrical extremity Sx	Local back pain May refer	Over the soft tissue involved
Description	Deep ache Vague	Variable pain Numbness, weakness Pseudoclaudication	Feeling of "slip" or "let go" Chronic, persistent ache	Sore Tight
Behavior or symptoms (Sx)	↑ Sx with sit, sit ↔ stand, cough/sneeze; ↓ Sx with lying,	↑ Sx with extension (prolonged standing or	↑ Sx with trivial stress, especially extension Stiffness after	↑ Sx with stretch and resistance

(continued)

Table 6.4a *(continued)*

Nature	Discogenic	Spinal stenosis	Instability (spondylolisthesis)	Strain or sprain
	unloading, walk (unless shifted) Stiffness after prolonged rest	walking especially down hill)	prolonged rest	
History (Hx)	Minor incident Often able to continue activity Notice initial or ↑ Sx after prolonged rest Difficult to straighten	Insidious onset Slow progression of Sx May have acute ↑ Sx due to secondary factor	Chronic Sx May have Hx of awkward lift or hyperextension injury (gymnastics)	Stretch injury
Objective exam	Posture may be shifted ↑ Sx with flexion Extension stiff	May have minimal findings with active motion Extension diminished and can be painful May have neurological findings	Excessive tenderness to palpation Possible step off at level of instability With progression of condition, has a hesitation in flexion at 30-40° With extension lumbar spine hinges at one segment May have neurological findings	Soft tissue thickening No neurological findings
Treatment (Rx)	Decrease compression load on disc Rotation is initially unloaded and gentle Intermittent traction Mobilization (especially central) McKenzie protocol	ADL instruction with neutral or slightly flexed posture (allowing > volume of spinal canal) Abdominal muscle strengtening/ stability Binder Mobilization— depending on probable extent and location of stenosis Intermittent lumbar traction Lumbar rotation	Improve trunk muscle strength regionally and segmentally Avoid postures/ activities that cause Sx External support Recognize that another level may be cause of Sx and may benefit from Rx with protection of the instability	Soft tissue Trigger points Posture correction Acute: rest and support

Table 6.4b Differential Diagnosis

Nature	Acute nerve root	Chronic nerve root	Acute facet	Chronic facet
Area	Dermatomal (patchy) Distal Sx > proximal	Dermatomal Distal Sx ≤ proximal	Unilateral pain Over joint May spread	Unilateral pain Over joint May spread
Description	Severe Sharp Shooting	Ache Shooting	Sharp	Less sharp
Behavior or symptoms (Sx)	Severely limits activity	Minimal limitation of activity	↑ Sx with stretch or compression of joint	↑ Sx with stretch of joint
History (Hx)	Can begin with ache proximally Sx more severe distally	Can give Hx of past acute never root Can be gradual, insidious onset	Flexion/rotation injury with local pain immediately May be difficult to straighten	Can give Hx of past acute facet Never entirely Sx free
Objective exam	Very limited motion Can be latent Neurological findings corresponding to the level of the problem may be diminished	May have minor neurological deficits Stiff at segment of problem	Local tenderness to palpation over affected joint	Palpation findings (stiff, thick, tender) over affected joint
Treatment (Rx)	In position of comfort Static traction Rotation toward side of Sx for lumbar Modalities at spine, optional distally	Local rotation Traction at that level	Rotation Traction at that level	Local rotation Traction at that level

PART
III

UPPER
EXTREMITY

SHOULDER JOINT COMPLEX

The shoulder joint complex consists of many different types of joints. Included in this myriad of joints are the glenohumeral joint, sternoclavicular joint, acromioclavicular joint, and scapulothoracic joint. It is imperative to assess the movements of each of these joints when evaluating the arthrokinematics of the shoulder. For this reason, arthrokinematics are listed after the description of all individual joints. The shoulder complex has a great range of motion and is a common area for upper-extremity pathology of various kinds. As always, any shoulder assessment should also include a cervical spine assessment.

GLENOHUMERAL JOINT (GH)

The glenohumeral joint is the true shoulder joint. Its anatomical make-up allows for multiplanar motion, important for functional use of the hand (figure 7.1).

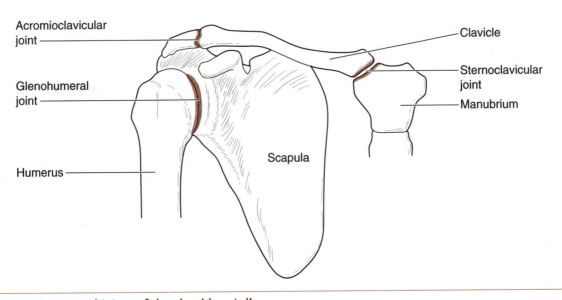

Figure 7.1 Bones and joints of the shoulder girdle.

Articulation

Convex humeral head and the concave glenoid fossa.

Type of Joint

Diarthrosis; spheroidal; ball and socket joint.

Degrees of Freedom

- Flexion and extension in sagittal plane about a coronal axis through the humeral head.
- Abduction and adduction in coronal plane about a sagittal axis through the humeral head.
- Internal and external rotation in transverse plane about a longitudinal axis through the humeral head.

In addition, circumduction and horizontal abduction and adduction occur at the glenohumeral joint.

Active Range of Motion

Flexion: 0-180°

Extension: 0-60°

Abduction: 0-180°

Internal rotation: 0-70°

External rotation: 0-90°

Horizontal adduction: 0-40°

Horizontal abduction: 0-90°

Accessory Movement

Anterior Glide

Assessment: External rotation, extension, horizontal abduction. You can assess anterior glide in a variety of abducted positions.

Patient: Prone with arm relaxed.

Operator position: Place one hand on the posterior humerus and the other hand supports the forearm

Mobilizing force: Apply an anterior force, bringing the humerus anterior in relation to the glenoid.

Lateral Distraction

Assessment: General hypomobility.

Patient: Supine with the arm relaxed.

Operator position: With both hands, grasp the proximal humeral head.

Mobilizing force: Apply a mobilizing force laterally.

Posterior Glide

Assessment: Internal rotation, flexion, horizontal adduction (figure 7.2).

Patient: Supine with arm shoulder in resting position and involved arm over side of table.

Operator position: Grasp the proximal humeral head with both hands; stabilize the scapula.

Mobilizing force: Apply a posterior force, bringing the humerus posterior in relation to the glenoid.

Figure 7.2 Posterior glide of the glenohumeral joint.

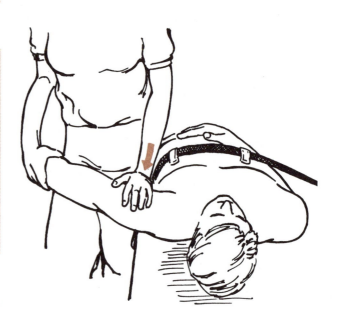

Inferior Glide

Assessment: Inferior glide of the humerus (figure 7.3).

Patient: Supine.

Operator position: Place the mobilizing hand on the superior humerus, distal to the acromion; place the other hand on the volar surface very close to the axilla.

Mobilizing force: Apply an inferior force to the humeral head.

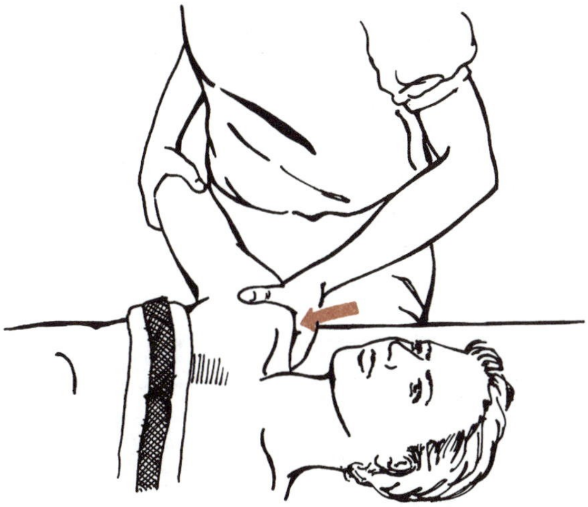

Figure 7.3 Inferior glide with shoulder near 90°.

End-Feel

Flexion: *Firm* from tissue stretch; posterior and inferior capsules become tight.

Extension: *Firm* from tissue stretch; anterior capsule becomes tight.

Abduction: *Hard* from humerus contacting acromial arch; *firm* from tissue stretch; inferior capsule becomes tight.

Adduction: *Soft* from tissue approximation.

Internal rotation: *Firm* from tissue stretch; posterior capsule, infraspinatus, or teres minor becomes tight.

External rotation: *Firm* from tissue stretch; anterior and inferior capsules become tight.

Horizontal abduction: *Firm* from tissue stretch; anterior capsule becomes tight.

Horizontal adduction: *Soft* from tissue approximation or *firm* from tissue stretch; posterior capsule, infraspinatus, or teres minor becomes tight.

Capsular Pattern

Restriction in external rotation followed by abduction followed by internal rotation.

Close-Packed Position

Maximal shoulder abduction and external rotation

Internal rotation and extension

Loose-Packed Position

60° abduction, 30° horizontal adduction.

Stability

Glenohumeral Ligaments

The glenohumeral ligaments attach to the upper and anterior edge of glenoid to anterior and inferior humeral head (figure 7.4).

Superior: provides anterior stabilization of humerus at 60° of elevation.

Middle: provides anterior stabilization of humerus at 90° of elevation.

Inferior: provides anterior stabilization of humerus at 120° of elevation.

All portions of the ligament are taut in humeral external rotation and abduction.

Coracohumeral Ligament

From the root of the coracoid process, the coracohumeral ligament passes laterally and downward to the greater tubercle blending with the supraspinatus tendon. It checks external rotation and extension and strengthens the superior capsule.

Capsule

The capsule attaches medially around glenoid fossa proximal to the labrum, extends to the root of the coracoid process enclosing the proximal attachment of the long head of the biceps, and attaches laterally at the anatomical neck of the humerus, extending a sleeve along the bicipital groove.

Glenoid Labrum

This is a fibrocartilaginous rim that deepens the glenoid cavity and makes the joint articulation more congruent.

Transverse Humeral Ligament

Bridges the gap between the greater and lesser trochanter and strengthens the shoulder capsule.

Coracoacromial Ligament

Blends with the trapezius and deltoid to form a roof over the humeral head; closes the coracoacromial arch, prevents superior dislocation of the humeral head.

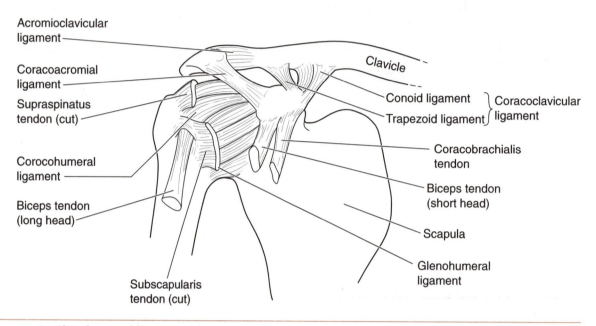

Figure 7.4 Glenohumeral ligaments.

Special Tests

Thoracic Outlet (Inlet) Syndrome Tests

Three-Minute Elevated Arm Test

Assessment: Possible neurovascular compression in the area of the thoracic outlet and inlet.

Patient: Sitting or standing.

Test position: Arms are abducted to 90° and elbows flexed to 90°. Patient is to maintain this position for three full minutes while opening and closing fists slowly.

Operator: Monitor symptoms.

Positive finding: Inability to complete task or onset of symptoms.

Pectoralis Minor Syndrome Test

Assessment: Constriction of middle portion of axillary artery by pectoralis minor.

Patient: Sitting.

Test position: Hyperabduction of the shoulder girdle. Have the patient hold breath and rotate neck away from arm.

Operator: Monitor pulse.

Positive finding: Diminution or disappearance of pulse or reproduction of neurological symptoms.

Costoclavicular Syndrome Test

Assessment: Compression of subclavian artery by the first rib and clavicle (figure 7.5).

Patient: Sitting.

Test position: Patient depresses and retracts arms from relaxed position.

Operator: Monitor radial pulse.

Positive finding: Diminution or disappearance of pulse or reproduction of neurological symptoms.

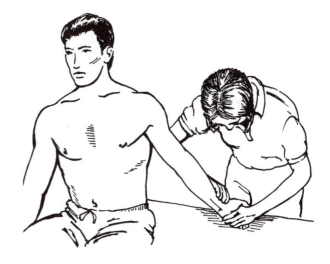

Figure 7.5 Costoclavicular syndrome test.

Biceps Brachii Test (Speed's Test)

Assessment: Biceps tendinitis (figure 7.6).

Patient: Supine.

Test position: 90° shoulder flexion and slight external rotation, elbow extension, forearm supination.

Operator: Apply resistance down into shoulder extension. (Can be done throughout shoulder range.)

Positive finding: Localized pain over biceps tendon origin.

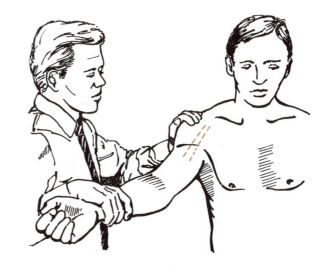

Figure 7.6 Biceps brachii test (Speed's test).

Adson Maneuver

Assessment: Scaleni compression of subclavian artery.

Patient: Sitting.

Test position: Abduct, retract, and externally rotate arm. Turn patient's head and hyperextend neck toward test side. Patient takes a deep breath.

Operator: Monitor radial pulse.

Positive finding: Diminution or disappearance of pulse or reproduction of neurological symptoms.

SECTION 7

Shoulder

Hawkins-Kennedy Impingement Test

Assessment: Impingement of supraspinatus tendon (figure 7.7, *a* and *b*).

Patient: Sitting.

Test position: Abduction and internal rotation of the arm in scapular plane.

Operator: Stabilize elbow and push down on wrist into more internal rotation.

Positive finding: Pain and reproduction of symptoms.

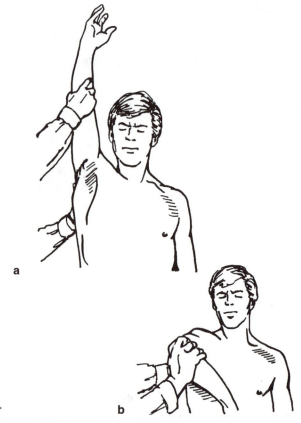

a

b

Figure 7.7 Hawkins-Kennedy impingement test.

Supraspinatus Impingement Test

Assessment: Supraspinatus tendinitis (figure 7.8).

Patient: Standing.

Test position: Shoulder elevated to 90° in scaption plane with full shoulder internal rotation.

Operator: Apply resistance to abduction.

Positive finding: Weakness or pain in the supraspinatus region.

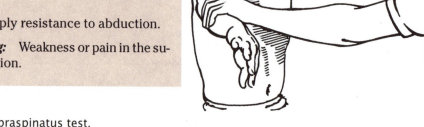

Figure 7.8 Supraspinatus test.

Transverse Humeral Ligament Test (for Subluxation of Long Head of Biceps)

Assessment: Test for integrity of transverse humeral ligament.

Patient: Sitting.

Test position: Elbow flexed.

Operator: Resist elbow flexion or passive external rotation.

Positive finding: Tendon subluxates medially, which may cause pain.

Rotator Cuff Rupture Test (Drop Arm Test)

Assessment: Rotator cuff rupture.

Patient: Sitting.

Test position: Arm is raised to 90° abduction.

Operator: Passively move arm to 90° abduction. Patient should hold arm as operator applies slight pressure.

Positive finding: Inability to hold arm in test position.

Crossover Impingement Test

Assessment: Impingement.

Patient: Sitting.

Test position: Horizontal adduction.

Operator: Apply overpressure into horizontal adduction.

Positive finding: Anterior pain—subscapularis, supraspinatus, long head of biceps; superior pain—acromioclavicular joint; posterior pain—infraspinatus, teres minor, posterior capsule.

Anterior Instability Tests

Glenohumeral Load and Shift

Assessment: Tests anterior stability of GH joint.

Patient: Sitting with arm supported on lap.

Test position: Place thumb across the posterior GH joint line and humeral head, web space between thumb and index finger across acromion, index finger across anterior GH joint line and humeral head, long finger over coracoid process.

Operator: Apply an anteromedial force to assess anterior stability.

Positive finding: Grading of GH translation (table 7.1).

Table 7.1 Glenohumeral Translation Grades

Trace	< 5 mm
I	5-10 mm, humeral head rides up glenoid slope, but not over the rim.
II	10-15 mm, humeral head rides up and over the glenoid rim, but spontaneously reduces when stress is removed.
III	> 15 mm, humeral head rides up and over the glenoid rim and remains dislocated when stress is removed.

SECTION 7

Shoulder

Relocation Test

Assessment: Anterior GH stability (figure 7.9).

Patient: Supine with arm over table.

Test position: Arm in 90° abduction and full external rotation.

Operator: Grasp the patient's forearm; grasp humeral head with other hand. Apply overpressure into external rotation. Next, apply a posterior force on the humeral head while applying overpressure into external rotation.

Positive finding: Reduction of pain with posterior force of humeral head and further pain-free external rotation motion.

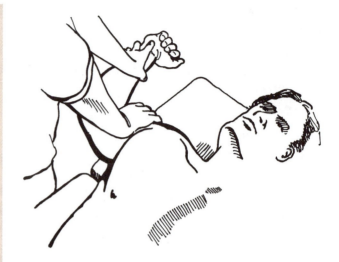

Figure 7.9   Relocation test.

Posterior Instability Tests

Glenohumeral Load and Shift

Assessment: Tests posterior stability of GH joint.

Patient: Sitting with arm supported on lap.

Test position: Place thumb across the posterior GH joint line and humeral head, web space between thumb and index finger across acromion, index finger across anterior GH joint line and humeral head, long finger over coracoid process.

Operator: Apply a posterolateral force to assess posterior stability.

Positive finding: Grading of GH translation (see table 7.1).

Posterior Glide: 90° Flexion and Internal Rotation

Assessment: Posterior instability.

Patient: Supine.

Test position: Shoulder in 90° of flexion and relaxed into internal rotation with elbow in 90° of flexion; allow posterior capsule to tighten.

Operator: Place one hand behind the shoulder across the posterior humeral head. The other hand applies an axial compressive force through the humerus.

Positive finding: Pain, apprehension, and instability.

Multidirectional Instability Test (Sulcus Sign)

Assessment: Multidirectional instability, especially inferior (figure 7.10).

Patient: Sitting with arm at side.

Test position: Arm in neutral and relaxed position.

Operator: Apply an inferior distraction force to arm.

Positive finding: Excessive inferior translation with sulcus defect at acromion.

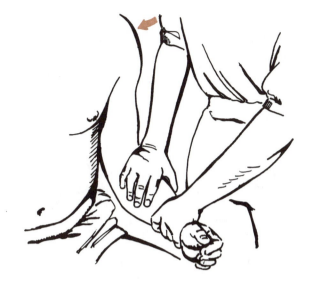

Figure 7.10 Sulcus sign test.

Glenoid Labrum Test (Clunk Test)

Assessment: Torn glenoid labrum (figure 7.11).

Patient: Supine.

Test position: Arm overhead in full abduction.

Operator: Place one hand on the posterior aspect of the humeral head; the other hand holds the humerus just proximal to the elbow. Apply an anterior force to the humeral head while applying compression and rotation to the humerus.

Positive finding: Pain, clunk, grinding, pseudolocking.

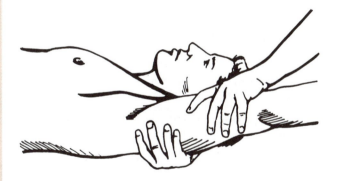

Figure 7.11 Clunk test.

Upper Limb Tension Tests

Radial Nerve

Assessment: Neural tension of radial nerve (figure 7.12).

Patient: Supine lying diagonally so scapula is off table edge, legs and other upper extremity resting on table in extended position.

Test position: Sequence of testing: shoulder (scapula) depression; elbow extension; forearm pronation and shoulder internal rotation simultaneously; wrist flexion, thumb flexion, and wrist ulnar deviation, shoulder abduction.

Operator: Face toward the feet of the patient. Your table-side hip is used to depress the shoulder girdle. The other hand grasps the patient's arm and progresses through the sequence. Sensitize above with lateral neck flexion.

Positive finding: Reproduction of symptoms.

Figure 7.12 Upper limb tension test of the radial nerve.

Median Nerve

Assessment: Neural tension of median nerve (figure 7.13).

Patient: Supine lying diagonal so scapula is off table edge.

Test position: Sequence of testing: shoulder depression; elbow extension; external rotation of shoulder; wrist and finger extension; and shoulder abduction.

Operator: Sensitize with lateral neck flexion.

Positive finding: Reproduction of symptoms.

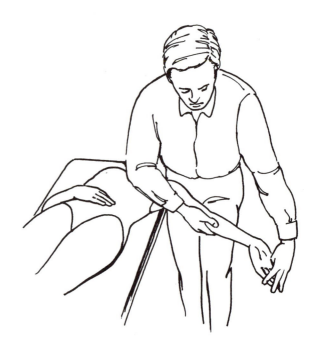

Figure 7.13 Upper limb tension test of the median nerve.

Ulnar Nerve

Assessment: Neural tension of ulnar nerve (figure 7.14).

Patient: Supine with shoulder near table edge and arm at side.

Test position: Sequence of testing: shoulder-girdle elevation blocked; patient's elbow at 90° flexion and elbow supported at therapist's anterior hip; wrist and finger extension; supinate or pronate; lateral rotation (optional); and slowly move into shoulder abduction.

Operator: Face patient; use one hand to depress the shoulder as the other hand maneuvers the arm. Sensitize above with lateral neck flexion.

Positive finding: Reproduction of symptoms.

Figure 7.14 Upper limb tension test of the ulnar nerve.

SECTION
7

Shoulder

Quadrant Test

Assessment: To assess articulating surface of the GH joint (figure 7.15).

Patient: Supine.

Test position: The operator's caudal arm slides palmar surface up, under the scapula so that the fingers can drape over the upper trapezius. The operator then grasps the elbow with the other hand, keeping the elbow flexed.

Operator: Abduct the shoulder through an arc of motion within the GH joint.

Positive finding: Pain, crepitus, limitation of range of motion.

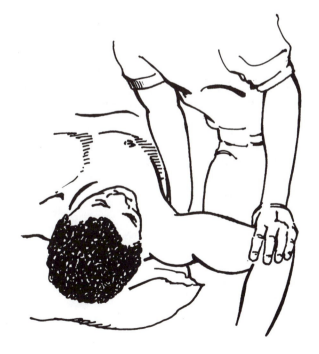

Figure 7.15 Maitland's quadrant test.

Locking Test

Assessment: Glenohumeral dysfunction.

Patient: Supine.

Test position: The operator's caudal arm slides palmar surface up, under the scapula so that the fingers can drape over the upper trapezius. The operator then grasps the elbow with the other hand, keeping the elbow flexed.

Operator: Move the arm from abduction toward full flexion, until it reaches a stopping point where it becomes locked.

Positive finding: Pain, joint grinding, reproduction of symptoms.

STERNOCLAVICULAR JOINT (SC)

Articulation

Clavicle articulates with manubrium of the sternum and cartilage of first rib; clavicle is convex cephalocaudally and concave anteroposteriorly; manubrium and first costal cartilage are concave cephalocaudally and convex anteroposteriorly. An articular disc divides the joint into two separate compartments.

Type of Joint

Diarthrodial, sellar joint.

Degrees of Freedom

- Elevation and depression in a coronal plane about a sagittal axis passes through the medial end of the clavicle.
- Protraction and retraction in a sagittal plane about a vertical axis passes longitudinally through the manubrium.
- Rotation in a transverse plane about a longitudinal axis passes lengthwise through the clavicle.

Active Range of Motion

None.

Accessory Movement

Ventral Glide

Assessment: To assess clavicular protraction. (figure 7.16).

Patient: Supine.

Operator position: Grasp the clavicle with your fingers superiorly and thumb inferiorly around the clavicle.

Mobilizing force: Lift the clavicle in an anterior direction.

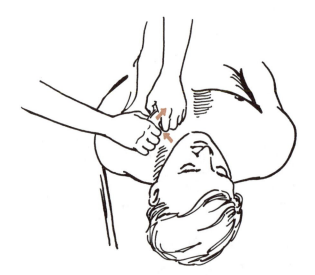

Figure 7.16 Ventral glide of the sternoclavicular joint.

Dorsal Glide

Assessment: To assess clavicular retraction

Patient: Supine.

Operator position: Place your thumb on the anterior surface of the proximal end of the clavicle; place the middle phalanx along the caudal surface of the clavicle to support the thumb.

Mobilizing force: Push with your thumb in a posterior direction.

Superior Glide

Assessment: To assess clavicular depression.

Patient: Supine.

Operator position: Grasp the clavicle with your fingers and thumbs.

Mobilizing force: Apply a superior glide to the clavicle.

Inferior Glide

Assessment: To assess clavicular elevation.

Patient: Supine.

Operator position: Grasp the clavicle with your fingers and thumbs.

Mobilizing force: Apply a caudal glide to the clavicle.

End-Feel

Not applicable; no active range of motion.

Capsular Pattern

Pain only at extremes of range.

Close-Packed Position

Maximal shoulder elevation.

Loose-Packed Position

Shoulder in anatomical neutral.

Stability

Articular Disc

Attached to clavicle and first costal cartilage; prevents medial dislocation of the clavicle.

Interclavicular Ligament

Extends from one clavicle to another, limits depression of the clavicle; supports weight of the upper extremity.

Costoclavicular Ligament

Extends from the inferior surface of the medial end of the clavicle to the first rib; limits protraction, retraction, and elevation of the clavicle.

Anterior Ligament

Extends from anterior manubrium of sternum to anteromedial end of the clavicle; reinforces the joint capsule and limits the posterior movement of the medial end of the clavicle associated with retraction and protraction.

Posterior SC Ligament

Extends from posterior sternum to posterior medial end of clavicle; reinforces the joint capsule and limits the anterior movement of the medial end of the clavicle associated with retraction and protraction.

Special Tests

No special tests for this joint.

ACROMIOCLAVICULAR JOINT (AC)

Articulation

Acromion of the scapula and distal end of the clavicle. Concave acromion and convex clavicular joint. Fibrocartilaginous disc separates articulation.

Type of Joint

Diarthrodial, plane synovial joint.

SECTION
7

Shoulder

Degrees of Freedom

- Tipping in sagittal plane about a coronal axis that passes lengthwise through the clavicle.
- Abduction and adduction in transverse plane about a longitudinal axis that passes through the lateral end of the clavicle.

Accessory Movement

Anterior Glide

Assessment: To assess anterior motion of the distal clavicle (figure 7.17).

Patient: Sitting with arm relaxed at the side; can also be done supine.

Operator position: Stand behind the patient and stabilize the acromion process with the fingers of your lateral hand. The thumb of your other hand is placed posteriorly on the clavicle.

Mobilizing force: Push the clavicle anteriorly with the thumbs.

Figure 7.17 Anterior glide of the acromioclavicular joint.

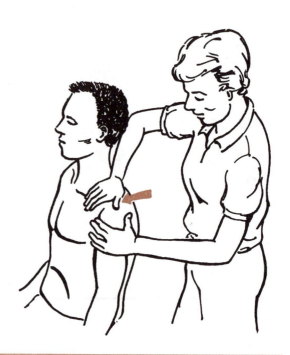

- Upward and downward rotation in coronal plane about a sagittal axis that passes through the lateral end of the clavicle.

Active Range of Motion

None.

End-Feel

Not applicable; no active range of motion.

Capsular Pattern

Pain at extremes of range of horizontal adduction.

Close-Packed Position

Arm abducted to 90°.

Loose-Packed Position

Shoulder in anatomical neutral.

Stability

Coracoclavicular Ligament

Holds clavicle to the coracoid process; the trapezoid runs in an anterolateral direction and pulls

Posterior Glide

Assessment: To assess posterior motion of the distal clavicle.

Patient: Sitting with arm relaxed at the side; can also be done supine.

Operator position: Stand behind the patient and stabilize the acromion process with the fingers of your lateral hand. The thumb of your other hand is placed posteriorly on the clavicle.

Mobilizing force: Push the clavicle anteriorly with the thumbs.

clavicle into backward rotation, also checks overriding or lateral movement of the clavicle on the acromion. The conoid portion runs in a posteromedial direction and checks superior movement of the clavicle on the acromion, and also prevents excessive widening of the scapuloclavicular angle.

Acromioclavicular Ligament

Runs from acromion process to the clavicle; maintains apposition of the joint, prevents posterior dislocation of the clavicle on the acromion.

Coracoacromial Ligament

Blends with the trapezius and deltoid to form a roof over the humeral head; closes the coracoacromial arch, prevents superior dislocation of the humeral head.

Special Tests

Acromioclavicular Shear Test

Assessment: Injury to the AC joint.

Patient: Sitting.

Test position: Arm relaxed and at side.

Operator: Cup hands together and compress hands together over AC joint, which creates a posterior-to-anterior glide of the joint.

Positive finding: Pain with compression.

Horizontal Adduction

Assessment: AC joint stability.

Patient: Sitting.

Test position: Patient horizontally adducts arm.

Operator: May apply slight overpressure.

Positive finding: Pain with movement.

SCAPULOTHORACIC JOINT (ST)

Articulation

Anterior surface of the scapula and posterior surface of the thorax.

Type of Joint

Not a true synovial joint; consists of the scapula and muscles covering posterior thoracic wall.

Degrees of Freedom

- Elevation and depression in the frontal plane either cranially (elevation) or caudally (depression).
- Abduction and adduction following the contour of the thoracic cage; abduction is movement away from the spine and adduction is movement toward it.
- Upward and downward rotation about a sagittal axis in which the inferior angle moves laterally for upward rotation and the inferior angle moves medially for downward rotation. Requires lateral and superior glide for upward rotation and medial and inferior glide for downward rotation.
- Anterior tilting about a coronal axis in which the inferior angle moves posteriorly.
- Winging of the scapula around the vertical axis.

Active Range of Motion

None.

Accessory Movement

Glides

Superior glide: To assess scapular elevation.

Inferior glide: To assess scapular depression.

Lateral glide: To assess abduction and protraction of the scapula.

Medial glide: To assess adduction and retraction of the scapula.

(continued)

SECTION 7

Shoulder

Combination of lateral and superior glide of the scapula: To assess abduction and upward rotation.

Combination of medial and inferior glide of the scapula: To assess adduction and downward rotation.

Patient: The glides listed can be assessed in the same position. Side-lying, patient faces the operator; the patient's arm is draped over the operator's inferior arm.

Operator position: Place the superior hand on the top of the shoulder over the acromion process. The inferior hand cups the scapula under the medial border and inferior angle.

Mobilizing force: Move the scapula in the direction to be assessed.

End-Feel

Not applicable; no active range of motion.

This joint also has no active movement associated with it and is not a true synovial joint, so information in the other categories does not apply to it.

ARTHROKINEMATICS

Flexion

ST: Scapula abducts and upwardly rotates in rhythm with GH movement to maintain muscle length-tension relationship.
AC: Clavicle glides superiorly.
SC: Clavicle glides inferiorly and laterally rotates.
GH: The humeral head glides inferiorly and posteriorly on the glenoid. Humerus undergoes lateral distraction while spinning internally.
Thoracic spine: Extends.

Extension

ST: Scapula adducts and downwardly rotates.
AC: Clavicle glides inferiorly and anteriorly.
SC: Clavicle glides superiorly and medially rotates.
GH: Humeral head glides superiorly and anteriorly.

Abduction

ST: Initial 30°—scapula is set against thorax 30° to full abduction; the scapula rotates upwardly and forward around rib cage.

AC: Clavicle glides inferiorly and anteriorly.
SC: Clavicle inferiorly glides from 90° to 120°; then clavicle rotates backward along its long axis.
GH: The humeral head glides inferiorly and rolls superiorly on the glenoid (the rotator cuff muscles counteract the superior rolling of the head of the humerus). Humerus spins into lateral rotation on its long axis as it elevates above 90°.
Thoracic spine: Extends.
Unilateral elevation: Upper thoracic spine must side-bend and rotate toward the side of elevation; lower thoracic side-bends away from side of motion.

Adduction

ST: The scapula rotates downwardly with minimal to no winging.
AC: Clavicle glides superiorly and posteriorly.
SC: Clavicle glides superiorly.
GH: Humeral head glides and rolls inferiorly on the glenoid.

External Rotation

ST: Scapula adducts as arm retracts.
SC: Clavicle rotates about long axis.
GH: Humeral head glides anteriorly.

Internal Rotation

ST: Scapula abducts as arm protracts.
SC: Clavicle rotates about long axis.
GH: Humeral head glides posteriorly.

Horizontal Abduction

ST: Scapula adducts.
AC: Distraction of the joint.
SC: Clavicle glides posteriorly.
GH: Anterior glide of humeral head.

Horizontal Adduction

ST: Scapula abducts.
AC: Compression of the joint.
SC: Clavicle glides anteriorly.
GH: Posterior glide of humeral head.

Scaption

Much the same as abduction, but the humerus elevates in the same plane as the scapula.

Circumduction

Combined movement of flexion, abduction, extension, and adduction.

Force Couples

1. Scapular upward rotation: upper and lower trapezius and the serratus anterior, with the center of rotation being within the scapula.

2. Humeral force couple: deltoid pulling up and the rotator cuff pulling down, with the center of rotation between the greater tuberosity and the deltoid tubercle.

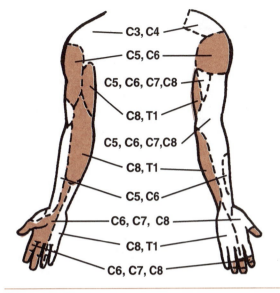

Figure 7.18 Cutaneous nerves.

NEUROLOGY

Table 7.2 Neurology

Nerve root	Reflex	Motor	Sensory
C5	Biceps reflex	Deltoid Biceps	Lateral arm
C6	Brachioradialis reflex	Wrist extensors Biceps	Lateral forearm
C7	Triceps reflex	Wrist flexors Finger extensors Triceps	Middle finger
C8	Abductor digiti minimi	Finger flexors Hand intrinsics	Medial forearm
T1		Hand intrinsics	Medial arm

Table 7.3 Peripheral Nerves

Nerve	Motor	Sensory
Axillary	Teres minor: shoulder external rotation	Lateral deltoid
Median	Abductor pollics brevis (ape hand)	DIP area of 1st finger
Radial	Extensor carpi radialis brevis (wrist drop)	Dorsal web space
Ulnar	Flexor carpi ulnaris (claw hand)	Dorsal and palmar 5th digit

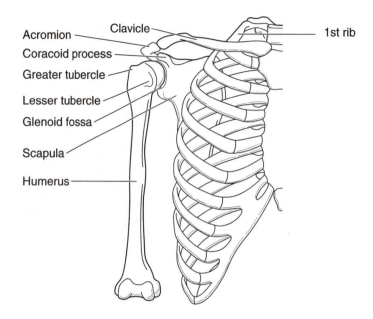

Acromion

Clavicle

1st rib

Coracoid process

Greater tubercle

Lesser tubercle

Glenoid fossa

Scapula

Humerus

a

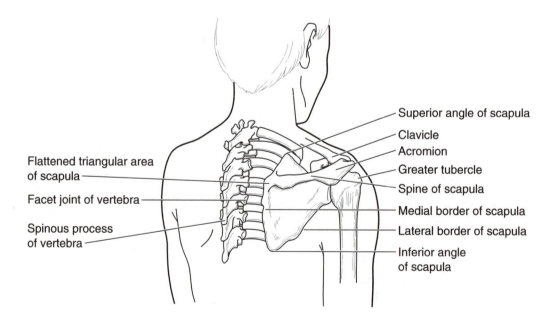

Superior angle of scapula

Clavicle

Acromion

Greater tubercle

Spine of scapula

Medial border of scapula

Lateral border of scapula

Inferior angle of scapula

Flattened triangular area of scapula

Facet joint of vertebra

Spinous process of vertebra

b

Figure 7.19 Shoulder bones: (*a*) anterior and (*b*) posterior views.

SURFACE PALPATION

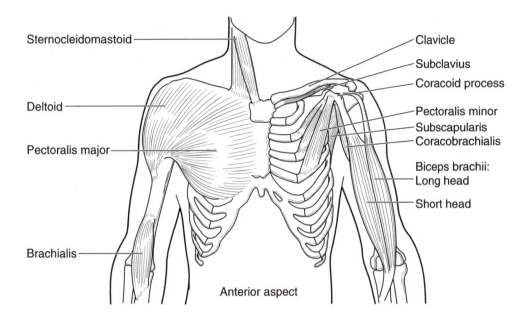

a

Sternocleidomastoid
Deltoid
Pectoralis major
Brachialis

Clavicle
Subclavius
Coracoid process
Pectoralis minor
Subscapularis
Coracobrachialis
Biceps brachii:
Long head
Short head

Anterior aspect

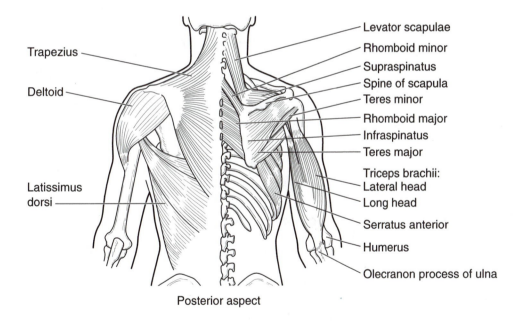

b

Trapezius
Deltoid
Latissimus dorsi

Levator scapulae
Rhomboid minor
Supraspinatus
Spine of scapula
Teres minor
Rhomboid major
Infraspinatus
Teres major
Triceps brachii:
Lateral head
Long head
Serratus anterior
Humerus
Olecranon process of ulna

Posterior aspect

Figure 7.20 Shoulder muscles.

SECTION 7

Shoulder

Anterior

Suprasternal notch

Sternum (manubrium)

Costal cartilage

Xiphoid process

SC joint

Clavicle

AC joint

Humerus (greater tubercle, lesser tubercle, bicipital groove)

Coracoid process

First rib

Posterior (Levels)

Scapula (acromion process, spine, medial border, inferior angle)

Spinous process of lower cervical and thoracic spine

Supraclavicular fossa

Infraclavicular fossa

MUSCLE ORIGIN AND INSERTION

Table 7.4   Muscle Origin and Insertion

Muscle	Origin and insertion
Biceps	Short head: apex of coracoid process of scapula. Long head: Supraglenoid tubercle *to* radial tuberosity and aponeurosis of biceps brachii
Coracobrachialis	Coracoid process of scapula *to* medial surface adjacent to deltoid tuberosity
Deltoid	Lateral third of clavicle, acromion process, spine of scapula *to* deltoid tubercle on humerus
Infraspinatus	Infraspionous fossa *to* greater tubercle on humerus
Latissimus dorsi	Spinous process of thoracic vertebrae 6-12 and lumbar 1-5 lower 3-4 ribs, iliac crest, inferior angle of scapula *to* intertubercle groove on humerus
Levator scapulae	Transverse process of cervical vertebrae 1-4 *to* superior angle of scapula
Rhomboids	Spinous process of C7 and T1-T5 *to* medial border of scapula
Serratus anterior	Ribs 1-8 *to* underside of scapula along medial border
Subscapularis	Subscapular fossa of scapula *to* lesser tubercle on humerus
Supraspinatus	Supraspinous fossa *to* superior facet of greater tubercle of humerus
Trapezius	Upper: external occipital protuberance, medial one third of superior nuchal line, ligamentum nuchae, spinous process of C7 Middle: spinous processes of T1-T5 Lower: spinous processes of T6-T12

Table 7.4 *(continued)*

Muscle	Origin and insertion
	Upper: lateral one third of clavicle and acromion process Middle: medial margin of acromion and superior lip of spine of scapula Lower: tubercle at apex of spine of scapula
Teres major	Posterior surface of scapula at inferior angle *to* lesser tubercle of humerus
Teres minor	Lateral border of posterior scapula *to* greater tubercle on humerus

MUSCLE ACTION AND INNERVATION

Table 7.5 Muscle Action and Innervation

Action	Muscle involved	Nerve supply	Nerve root
Forward flexion	1. Deltoid (anterior fibers)	Axillary	C5, C6 (post. cord)
	2. Pectoralis major (clavicular fibers)	Lateral pectoral	C5, C6 (lat. cord)
	3. Coracobrachialis	Musculocutaneous	C5, C6, C7
	4. Biceps	Musculocutaneous	C5, C6, C7
Extension	1. Deltoid (posterior fibers)	Axillary	C5, C6 (post. cord)
	2. Teres major	Subscapular	C5, C6 (Post. cord)
	3. Teres minor	Axillary	C5, C6 (Post. cord)
	4. Latissimus dorsi	Thoracodorsal	C6, C7, C8
	5. Pectoralis major (clavicular fibers)	Lateral pectoral	C5, C6 (lat. cord)
	6. Triceps (long head)	Radial	C5, C6, C7, C8, T1
Horizontal adduction	1. Deltoid (posterior fibers)	Axillary	C5, C6 (post. cord)
	2. Pectoralis major (clavicular fibers)	Lateral pectoral	C5, C6 (lat. cord)
Horizontal adbuction	1. Deltoid (posterior fibers)	Axillary	C5, C6 (post. cord)
	2. Teres major	Subscapular	C5, C6 (post. cord)

(continued)

Table 7.5 *(continued)*

Action	Muscle involved	Nerve supply	Nerve root
	2. Teres major	Subscapular	C5, C6 (post. cord)
	3. Teres minor	Axillary	C5, C6 (post. cord)
	4. Infraspinatus	Suprascapular	C5, C6, (brachial plexus)
Abduction	1. Deltoid 2. Supraspinatus 3. Infraspinatus 4. Subscapularis 5. Teres minor	Axillary Suprascapular Suprascapular Suprascapular Axillary	C5, C6 C5, C6 C5, C6 C5, C6 C5, C6
Adduction	1. Pectoralis major 2. Latissimus dorsi 3. Teres minor	Lateral pectoal Thoracodorsal Axillary	C5, C6 C6, C7, C8 C5, C6
Internal rotation	1. Pectoralis major 2. Latissimus dorsi 3. Teres minor 4. Deltoid (anterior fibers) 5. Subscapularis	Lateral pectoral Thoracodorsal Axillary Axillary Suprascapular	C5, C6 C6, C7, C8 C5, C6 C5, C6 C5, C6
External rotation	1. Infraspinatus 2. Teres minor 3. Deltoid	Suprascapular Axillary Axillary	C5, C6 C5, C6 C5, C6
Elevation of scapula	1. Trapezius (upper fibers) 2. Levator scapulae 3. Rhomboid major 4. Rhomboid minor	Accessory C3, C4 nerve roots C3, C4 nerve roots Dorsal scapular Dorsal scapular Dorsal scapular	Cranial nerve XI C3, C4 C5 C5
Depression of scapula	1. Serratus anterior 2, Pectoralis major 3. Latissimus dorsi 4. Pectoralis minor 5. Trapizius (low fibers)	Long thoracic Lateral pectoral Thoracodorsal Medial pectoral Accessory C3, C4 nerve roots	C5, C6 C5, C6 C6, C7, C8 C8, T1 Cranial nerve XI
Scapular protraction	1. Serratus anterior 2. Pectoralis major 3. Latissimus dorsi 4. Pectoralis minor	Long thoracic Lateral pectoral Thoracodorsal Medial pectoral	C5, C6 C5, C6 C6, C7, C8 C8, T1

The Clinical Orthopedic Assessment Guide

Table 7.5 (continued)

Action	Muscle involved	Nerve supply	Nerve root
Scapular retraction	1. Trapezius	Accessory	Cranial nerve XI
	2. Rhomboid major	Dorsal scapular	C5
	3. Rhomboid minor	Dorsal scapular	C5
Scapular upward rotation	1. Trapezius (upper fibers)	Accessory	Cranial nerve XI C3, C4, nerve roots
	2. Serratus anterior	Long thoracic	C5, C6
Scapular downward rotation	1. Levator scapulae	C3, C4 nerve roots Dorsal scapular	C3, C4
	2. Rhomboid major	Dorsal scapular	C5
	3. Rhomboid minor	Dorsal scapular	C5
	4. Pectoralis minor	Medial pectoral	C8, T1

DIFFERENTIAL DIAGNOSIS

Table 7.6   Differential Diagnosis

Disorder	Description	Onset	Symptoms	Signs	Special tests
AC sprain	Separation of the acromioclavicular joint, ligamentous tear	Fall on outstretched arm or on tip of shoulder	Local pain with elevation > 90°; pain with horizontal adduction	Local swelling; increased joint play at AC joint; deformity of AC joint	Horizontal adduction
Bicipital tendinitis	Inflammation of the biceps tendon at its origin	Repetitive use of biceps muscle	Local anterior tenderness	Guarding upper extremity involved; shoulder flexion causes pain	Speed's test
Supraspinatus tendinitis	Inflammation of the supraspinatus tendon	Chronic, repetitive overhead activities	Pain with overhead activity, pain at night	Painful arc (60-120°); point tender	Impingement tests

(continued)

Table 7.6 *(continued)*

Disorder	Description	Onset	Symptoms	Signs	Special tests
Frozen shoulder	Adhesive capsulitis	Insidious	Diffuse pain and stiffness	Capsular limitation in AROM: decreases joint play	Accessory movements limited
Glenohumeral dislocation	Humerus completely separates from glenoid	Fall on out-stretched arm; forced abduction and external rotation	Severe pain; instability	Obvious deformity, excessive glide	Glenohumeral load and shift, relocation test
Rotator cuff tear	Muscle tear in one of the rotator cuff muscles	Acute trauma; degenerative changes	Pain with rotation or abduction; unable to apply pressure to shoulder	Unable to resist abduction; atrophy	Drop arm test
Subacromial bursitis	Inflammation of subacromial bursa	Overuse; repetitive overhead movement	Severe local tenderness	Lateral swelling; painful arc; noncapsular pattern of ROM; empty end-feel	Tenderness to palpation
Thoracic outlet syndrome	Compression of neurovas-cular bundle as it passes through the thoracic inlet	Insidious	Pain and sensory changes	Poor posture with rounded shoulders; weakness in C8-T1 musculature	Adson tests, costoclavicular test, pectoralis minor test

SECTION 8

ELBOW AND FOREARM

Elbow and forearm motions are important for positioning of the hand for function. The elbow and forearm consist of the humeroulnar, humeroradial, and superior and inferior radioulnar joints. The complex arthrokinematics are key in function of this upper extremity.

HUMEROULNAR JOINT (HU)

Articulation
Convex trochlea of the humerus and concave trochlear notch on proximal ulna.

Type of Joint
Diarthrodial hinge joint.

Degrees of Freedom
- Flexion and extension in sagittal plane about a coronal axis through the humeral epicondyles.
- Abduction and adduction in frontal plane about a sagittal axis through the humeral epicondyles

Active Range of Motion
Elbow flexion: 0-150°.

Elbow extension: 0-10° of hyperextension

Accessory Movement

Joint Distraction

Assessment: To assess general mobility.

Patient: Supine, elbow slightly bent over the edge; may use stabilization belt to hold humerus down.

Operator position: Place interlaced fingers over proximal ulna on volar surface.

Mobilizing force: Apply force against ulna at a 45° angle to shaft.

Medial Glide of Ulna

Assessment: To assess elbow extension.

Patient: Supine with elbow flexed to 70° and forearm supinated.

Operator position: Stand beside patient; one hand stabilizes distal humerus and other hand holds the forearm, proximally near the elbow.

(continued)

93

> **Mobilizing force:** Apply a medially directed force.

Lateral Glide of Ulna

Assessment: To assess elbow flexion.

Patient: Supine with elbow flexed to 70° and forearm supinated.

Operator position: Stand beside patient; one hand stabilizes distal humerus and other hand holds the forearm.

Mobilizing force: Apply a lateral force.

End-Feel

Flexion: *Soft* due to muscle bulk of anterior forearm and anterior arm

Flexion: *Hard* due to minimal muscle bulk; coronoid process contacts coronoid fossa

Flexion: *Firm* due to tension in the posterior joint capsule and the triceps

Extension: *Hard* due to olecranon process contacting the olecranon fossa

Extension: *Firm* due to tensions in the anterior joint capsule, collateral ligaments, and biceps

Capsular Pattern

Greater limitation in flexion than extension.

Close-Packed Position

Full extension with forearm supination.

Loose-Packed Position

70-90° of flexion, 10° of supination.

Stability

Medial (Ulnar) Collateral Ligament

Anterior band: runs from anterior aspect of medial epicondyle to medial coronoid process.

Posterior band: runs from posterior medial epicondyle to medial olecranon (figure 8.1, *a* and *b*).

Oblique band: Connects anterior and posterior band; restricts medial angulation of the ulna on the humerus.

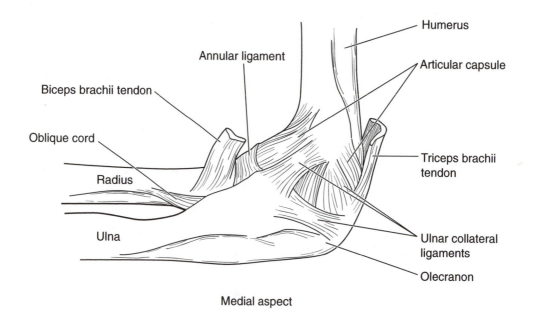

Medial aspect

a

Figure 8.1 Elbow ligaments.

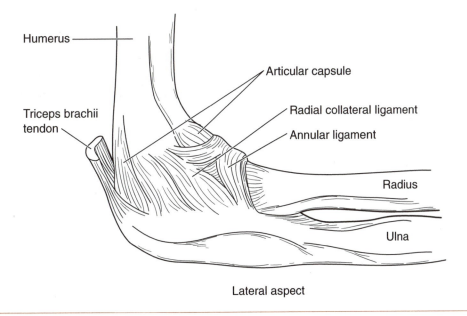

Humerus

Triceps brachii tendon

Articular capsule

Radial collateral ligament

Annular ligament

Radius

Ulna

b

Lateral aspect

Figure 8.1 Elbow ligaments.

Capsule

Encloses humeroulnar joint, radiohumeral joint, and proximal radioulnar joint. Anteriorly and above, it is attached to the humerus along the upper margins of the coronoid and radial fossae and to the front of the medial and lateral epicondyles; below, it is attached to the margin of the coronoid process of the ulna and to the annular ligament. Posteriorly and above, it is attached to the margins of the olecranon fossa of the humerus; below, it is attached to the upper margin and sides of the olecranon process of the ulna and to the annular ligament.

Special Tests

Medial Collateral Stress Test

Assessment: Integrity of medial collateral ligament.

Patient: Sitting.

Test position: Arm is relaxed and cradled in tester's hands.

Operator: Apply a valgus test to stress the medial collateral ligament.

Positive finding: Abnormal gapping of the joint when stress is applied.

Tinel's Sign

Assessment: Involvement of the ulnar nerve.

Patient: Sitting.

Test position: Arm relaxed and supported by table.

Operator: Tap along the ulnar nerve where it travels along the groove between the olecranon and the medial epicondyle.

Positive finding: Tingling sensation that reproduces the patient's symptoms.

Little Leaguer's Elbow (Golfer's Elbow) Test

Assessment: Medial epicondylitis.

Patient: Sitting.

Test position: Arm extended and in supination.

Operator: Resist wrist flexion.

Positive finding: Pain along medial epicondyle.

Pinch Test

Assessment: Entrapment of anterior interosseous nerve as it passes between the two heads of the pronator teres.

Patient: Sitting.

Test position: Patient pinches tip of index finger and thumb; tip-to-tip pinch.

Operator: Instruct patient on test position.

Positive finding: Abnormal tip-to-tip pinch; has to pinch pulp to pulp.

HUMERORADIAL JOINT (HR)

Articulation

Convex capitulum of humerus and concave radial head.

Type of Joint

Diarthrodial sellar joint.

Degrees of Freedom

- Flexion and extension in sagittal plane about a coronal axis.

Accessory Movement

Joint Traction

Assessment: To assess joint mobility (figure 8.2)

Patient: Supine or sitting.

Operator position: Place yourself between the treatment table and the patient's arm; stabilize the humerus with your superior hand. Grasp around the distal radius with your other hand.

Mobilizing force: Pull the radius along its long axis.

Figure 8.2 Joint traction during humeroradial articulation.

Elbow Quadrant Test

Assessment: Joint integrity.

Patient: Supine with arm relaxed.

Test position: Elbow with slight flexion.

Operator: One hand grasps distal forearm and the other hand is stabilizing distal humerus; take the forearm through small arc of motion.

Positive finding: Reproduction of symptoms and/or crepitus.

- Abduction and adduction in coronal plane about a sagittal axis.
- Internal and external rotation in transverse plane about a vertical axis.

Active Range of Motion

Active range of motion for the elbow is the same as for humeroulnar joint.

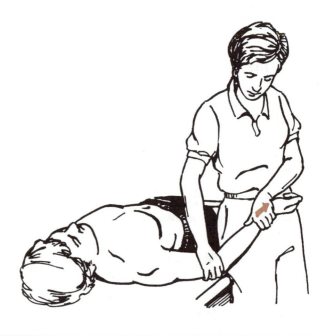

Dorsal/Volar Glide

Assessment: To assess flexion (volar glide) and extension (dorsal glide) (figure 8.3).

Patient: Supine with the elbow extended and supinated.

Operator position: Face the patient. Stabilize the humerus with your inside hand; place the palmar surface of your lateral hand on the volar aspect and your fingers on the dorsal aspect of the radial head.

Mobilizing force: Force the radial head dorsally with the palm of your hand or volarly with your fingers.

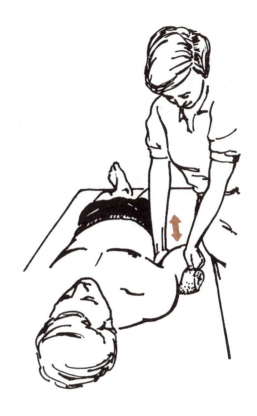

Figure 8.3 Dorsal/volar glide during humeroradial articulation.

End-Feel

Flexion: *Soft* due to soft tissue block or bony block in thin individual

Extension: *Hard* bony block (olecranon fossa meets olecranon)

Forearm pronation: *Firm* due to muscular end-feel

Forearm supination: *Firm* due to ligament

Capsular Pattern

Flexion, extension, supination, pronation.

Close-Packed Position

90° flexion and 5° supination.

Loose-Packed Position

70° flexion and 35° supination.

Stability

Lateral (radial) collateral ligament: attached by its apex to the lateral epicondyle of the humerus and by its base to the upper margin of the annular ligament.

Special Tests

Lateral Collateral Stress Test

Assessment: Integrity of lateral collateral ligament.

Patient: Sitting.

Test position: Arm is relaxed and cradled in tester's hands.

Operator: Apply a varus test to stress the lateral collateral ligament.

Positive finding: Abnormal gapping of the joint when stress is applied.

Tennis Elbow Test

Assessment: Lateral epicondylitis (figure 8.4).

Patient: Sitting.

Test position: Flex arm to approximately 70°, fist clenched and wrist extended.

Operator: Apply resistance against wrist extension.

Positive finding: Pain along lateral epicondyle.

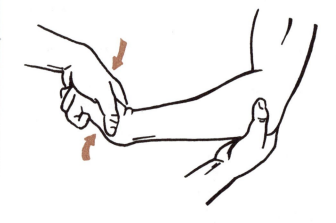

Figure 8.4 Tennis elbow test.

SUPERIOR RADIOULNAR JOINT (SRU)

Articulation

Concave radial notch and annular ligament and convex head of the radius.

Type of Joint

Diarthrodial pivot joint.

Accessory Movement

Degrees of Freedom

Supination and pronation in transverse plane about a vertical axis.

Active Range of Motion

Pronation: 0-80°
Supination: 0-80°

Dorsal/Volar Glide

Assessment: To assess pronation and supination (figure 8.5).

Patient: Sitting with elbow and forearm in resting position.

Operator position: Fixate the ulna with your medial hand around the medial aspect of the forearm; place your other hand around the head of the radius with fingers on the volar surface and the palm on the dorsal surface.

Mobilizing force: Force the radial head volarly by pushing with your palm, or dorsally by pulling with your fingers.

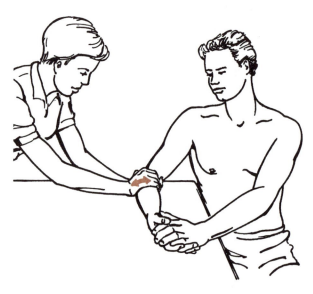

Figure 8.5 Dorsal/volar glide in the radioulnar joint.

End-Feel

Pronation: *Hard* (ulna contacts the radius)

Pronation: *Firm* (due to tension in the radioulnar ligament, supinator, interosseous membrane)

Supination: *Firm* (due to tension in the palmar radioulnar ligament of the distal radioulnar joint, interosseous membrane, pronator quadratus, quadrate ligament)

Capsular Pattern

Equal limitation of supination and pronation.

Close-Packed Position

Full pronation and full supination.

Loose-Packed Position

70° flexion, 35° supination.

Stability

Annular Ligament

Runs from anterior margin of the radial notch of the ulna around the radial head to the posterior margin of the radial notch, blends with capsule and lateral ligament; protects radial head and stabilizes radial head next to ulna.

Quadrate Ligament

Runs from inferior edge of radial notch to the neck of radius; reinforces inferior joint capsule and limits supination.

Oblique Cord

Runs from just inferior to radial notch on ulna to inferior portion of bicipital tuberosity of radius, resists distal displacement of the radius during pulling movements.

Interosseous Membrane

Runs between radius and ulna to provide transmission of forces.

Special Tests

No special tests for this joint.

DISTAL RADIOULNAR JOINT

Articulation

Convex ulnar head and concave ulnar notch of the radius.

Type of Joint

Diarthrodial pivot joint.

Degrees of Freedom

Supination and pronation in transverse plane about a vertical axis.

Active Range of Motion

Pronation: 0-80°

Supination: 0-80°

Accessory Movement

Dorsal/Volar Glide

Assessment: Pronation and supination.

Patient: Sitting with arm on the treatment table in resting position.

Operator position: Stabilize the distal ulna by placing the fingers of one hand on the dorsal surface and the thenar eminence and thumb on the volar surface. Place your other hand in the same manner around the distal radius.

Mobilizing force: Glide the distal radius dorsally or volarly parallel to the ulna.

End-Feel

Supination: *Firm* due to ligamentous restraints

Pronation: *Firm* due to muscular end-feel

Capsular Pattern

Full range with pain at extreme ranges.

Close-Packed Position

5° of supination.

Loose-Packed Position

10° of supination.

Stability

Interosseous Membrane

Consists of broad sheet of collagenous tissue.

Articular Disc

Proximally articulates with the ulnar head and distally with the lunate and triquetrum; holds radius and ulna together and helps to transfer forces.

Anterior Radioulnar Ligament

Attaches above ulnar head to ulnar notch and stabilizes anterior aspect of the joint.

Posterior radioulnar ligament: posterior aspect of head of ulna to posterior ulnar notch; stabilizes posterior aspect of the joint.

Special Tests

No special tests for this joint.

ARTHROKINEMATICS

Extension

HU: Ulna slides posteriorly, superiorly, and medially on the trochlea until the ulnar olecranon process enters the olecranon fossa; ulna deviates laterally and internally rotates.

HR: The concave surface of the radial head slides posteriorly on the capitulum.

Flexion

HU: The trochlear ridge of the ulna slides anteriorly, superiorly, laterally along the trochlear groove until the coronoid process reaches the floor of the coronoid fossa; ulna deviates medially and externally rotates.

HR: The rim of the radial head slides anteriorly in the capitulotrochlear groove to enter the radial fossa.

Pronation

SRU: The convex rim of the head of the radius spins posteriorly within the annular ligament and the concave radial notch. The ulnar head moves distally and dorsally.

Supination

SRU: The convex rim of the head of the radius spins anteriorly within the annular ligament and the concave radial notch. The ulnar head moves proximally and volarly.

NEUROLOGY

Table 8.1   Neurology

Nerve root	Reflex	Muscle°	Sensory
C5	Biceps	Biceps	Lateral arm
C6	Brachioradialis	Wrist extensors	Lateral forearm
C7	Triceps	Triceps	Index and middle finger
C8	Abductor digiti minimi	Thumb extensors	Medial forearm
T1	None	Finger abductors	Medial arm

Table 8.2   Pheripheral Nerves

Nerve	Motor	Sensory
Anterior interosseous	FPL, radial half of FDP, PQ	
Median	Abductor pollics brevis (ape hand)	DIP area of 1st finger
Radial	Extensor carpi radialis brevis (wrist drop)	Dorsal web space
Ulnar	Flexor carpi ulnaris (claw hand)	Dorsal/palmar 5th digit

SURFACE PALPATION

Anterior

Medial epicondyle

Lateral epicondyle

Medial supracondylar notch

Lateral supracondylar notch

Cubital fossa (lateral to medial): biceps tendon, brachial artery, median nerve

Brachioradialis, pronator teres, tendon of biceps, bicipital aponeurosis, brachialis

Median nerve

Cubital vein

Ulnar styloid

Radial styloid

Flexor carpi ulnaris

Palmaris longus

Flexor carpi radialis

Pronator teres

Posterior

Olecranon process

Olecranon fossa

Triceps tendon

Ulnar nerve in cubital tunnel

Head of the radius

Anconeus

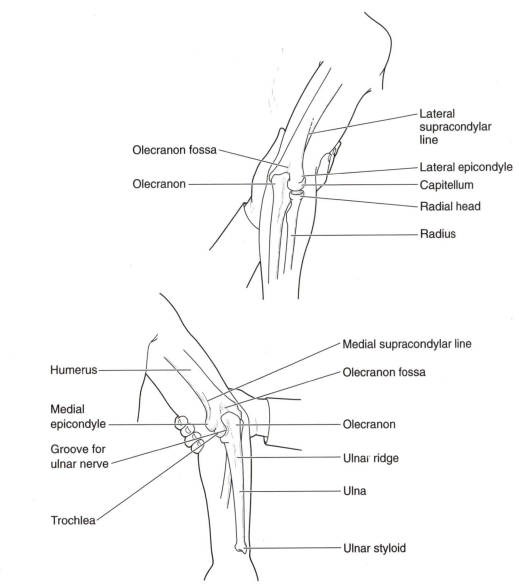

Figure 8.6 Palpation around the elbow: (*a*) posterolateral view and (*b*) posteromedial view.

MUSCLE ORIGIN AND INSERTION

Table 8.3 Muscle Origin and Insertion

Muscle	Origin and insertion
Anconeus	Lateral epicondyle of humerus *to* olecranon process on ulna
Biceps	Short head: apex of coracoid process of scapula Long head: supraglenoid tubercle *to* radial tuberosity and aponeurosis of biceps brachii
Brachialis	Anterior surface of lower humerus *to* coronoid process on ulna
Brachioradialis	Lateral supracondylar ridge of humerus *to* styloid process of radius
Extensor carpi radialis brevis	Lateral epicondyle of humerus *to* base of 3rd metacarpal
Extensor carpi radialis longus	Lateral supracondylar ridge of humerus *to* base of 2nd metacarpal
Extensor carpi ulnaris	Lateral epicondyle *to* base of 5th metacarpal
Flexor carpi ulnaris	Medial epicondyle *to* pisiform, hamate, base of 5th metacarpals
Flexor carpi radialis	Medial epicondyle *to* base of 2nd and 3rd metacarpals
Pronator quadratus	Distal anterior surface of ulna *to* distal anterior surface of radius
Pronator teres	Medial epicondyle and coronoid process on ulna *to* midway down on the lateral surface of radius
Supinator	Lateral epicondyle *to* upper, lateral side of radius
Triceps	Long head: infraglenoid tubercle of scapula Lateral head: lateral and posterior surfaces of proximal one-half of body of humerus, and lateral intermuscular septum Medial head: distal two-thirds of medial and posterior surfaces of humerus below the radial groove and from medial intermuscular septum *to* olecranon process and antebrachial fascia

MUSCLE ACTION AND INNERVATION

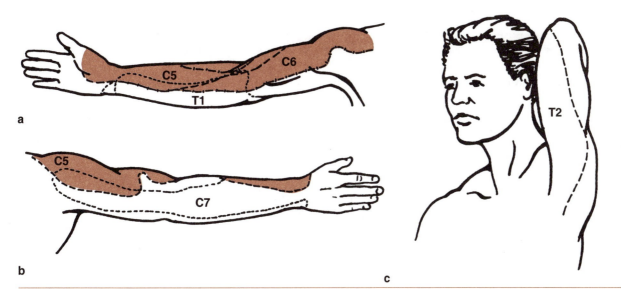

Figure 8.7 Dermatomes around the elbow.

Table 8.4 Muscle Action and Innervation

Action	Muscle involved	Nerve supply	Nerve root
Elbow flexion	Brachialis	Musculocutaneous	C5-C6 (C7)
	Biceps brachii	Musculocutaneous	C5-C6
	Brachioradialis	Radial	C5-C6
	Pronator teres	Median	C6-C7
	Flexor carpi ulnaris	Ulnar	C7-C8
Elbow extension	Triceps	Radial	C6-C8
	Anconeus	Radial	C7-C8
Forearm supination	Supinator	Radial	C5-C6
	Biceps brachii	Musculocutaneous	C5-C6
Forearm pronation	Pronator quadratus	Median	C5-C6
	Pronator teres	Median	C6-C7
	Flexor carpi radialis	Median	C6-C7
Wrist flexion	Flexor carpi radialis	Median	C6-C7
	Flexor carpi ulnaris	Ulnar	C7-C8
Wrist extension	Extensor carpi radialis longus	Radial	C6-C7
	Extensor carpi radialis brevis	Radial	C7-C8
	Extensor carpi ulnaris	Radial	C7-C8

SECTION

8

Elbow and Forearm

DIFFERENTIAL DIAGNOSIS

Table 8.5 Differential Diagnosis

Disorder	Description	Onset	Symptoms	Signs	Special tests
Lateral epicondylitis	Tendinitis of wrist extensor and comon origin on lateral epicondyle	Gradual onset usually involving repetitive wrist extention	Dull ache at rest; sharp pain with lifting activity	Resisted wrist extension causes pain; tenderness over lateral epicondyle	Tennis elbow test
Medial epicondylitis	Tendinitis of wrist flexors and comon origin on medial epicondyle	Gradual onset usually involving repetitive wrist flexion	Pain with wrist and finger flexion	Resisted wrist flexion; tenderness over medial epicondyle	
Olecranon bursitis	Inflammation of the olecranon bursa	Continuous pressure on olecranon; or acute trauma	Pain	Swelling; limited range of motion	
Biceps muscle rupture	Disruption of biceps from attachment (usually distal)	Quick forceful biceps contraction	Pain in posterior elbow	Discontinuity of biceps with bulge, loss of elbow flexion strength	
Reflex sympathetic dystrophy	Hypovascularity by increased sympathetic activity	Gradual after period of immobilization	Pain, swelling	Pitting edema, tenderness, skin changes, bluish skin, restricted movement	
Volkmann's ischemic contracture	Compartment syndrome resulting in decreased circulation and nerve disturbance	Following fracture or dislocation of the elbow	Painful fingers, paresthesia	Purple discoloration of hand, loss of radial pulse, clawed fingers	
Myositis ossificans	Ossification of soft tissue	Gradual onset from aggressive stress on soft tissue	Pain with range of motion	Hard mass in soft tissue area	X-ray confirmation

The Clinical Orthopedic Assessment Guide

SECTION 9

WRIST AND HAND

The wrist and hand make up 90% of the use of the upper extremity. The wrist and hand complex consists of the radiocarpal and midcarpal joints at the wrist and the carpometacarpal, metacarpal, and interphalangeal joints of the digits. Even though the shoulder, elbow, and wrist and hand are listed in separate sections, the clinician needs to assess the entire upper-extremity chain.

RADIOCARPAL JOINT

Articulation

A biconcave surface formed by the radial facet and radioulnar disc with a biconvex surface consisting of scaphoid, lunate, and triquetrum.

Type of Joint

Diarthrodial ellipsoid joint.

Degrees of Freedom

- Flexion and extension in sagittal plane about a coronal axis through the head of the capitate
- Ulnar and radial deviation in coronal plane about a sagittal axis through the capitate

Active Range of Motion (American Academy of Orthopedic Surgeons)

Flexion: 0-80°

Extension: 0-70°

Ulnar deviation: 0-30°

Radial deviation: 0-20°

Accessory Movement

Distraction

Assessment: To assess joint mobility of the wrist.

Patient: Sitting with forearm supported on table palm down, wrist over the edge.

Operator position: With one hand, grasp the wrist around the styloid process. The other hand grasps around the distal row of carpals; operator's forearms parallel to patient's forearms.

Mobilizing force: Pull in the direction of the long axis of the forearm.

Volar Glide

Assessment: To assess wrist extension (figure 9.1).

Patient: Sitting with forearm supported on table palm down, wrist over the edge.

Operator position: With one hand, grasp the wrist around the styloid process. The other hand grasps around the distal row of carpals.

Mobilizing force: Force is directed in a volar direction.

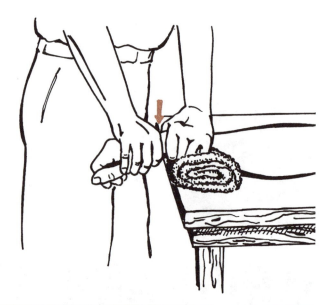

Figure 9.1 Volar glide in the mobilization of the wrist.

Ulnar Glide

Assessment: To assess radial deviation (figure 9.2).

Patient: Sitting with forearm supported on table palm down, wrist over the edge.

Operator position: With one hand, grasp the wrist around the styloid process. The other hand grasps around the distal row of carpals.

Mobilizing force: Force is directed toward the ulnar border, perpendicular to line of forearm.

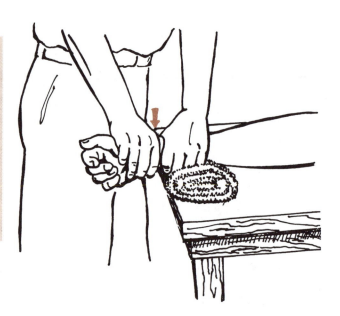

Figure 9.2 Ulnar glide in mobilization of the wrist.

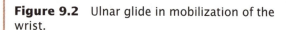

Dorsal Glide

Assessment: To assess wrist flexion.

Patient: Sitting with forearm supported on table palm down, wrist over the edge.

Operator position: With one hand, grasp the wrist around the styloid process. The other hand grasps around the distal row of carpals.

Mobilizing force: Force is directed in a dorsal direction.

Radial Glide

Assessment: To assess ulnar deviation.

Patient: Sitting with forearm resting on the table thumb-side up.

Operator position: With one hand, grasp the wrist around the styloid process. The other hand grasps around the distal row of carpals.

Mobilizing force: Force is directed toward radial border, perpendicular to line of forearm.

End-Feel

Flexion: *Firm* due to tension in dorsal radiocarpal ligament and dorsal joint capsule

Extension: *Firm* due to tension in palmar radiocarpal ligament and palmar joint capsule

Extension: *Hard* (radius contacts carpal bones)

Radial deviation: *Hard* (radial styloid contacts scaphoid)

Radial deviation: *Firm* due to tension in ulnar collateral ulnocarpal ligament and ulnar portion of the joint capsule

Ulnar deviation: *Firm* due to tension in radial collateral ligament and radial portion of the joint capsule

Capsular Pattern

Equal limitation of flexion and extension.

Close-Packed Position

Extension with radial deviation.

Loose-Packed Position

Loose-packed position is the resting position of the hand; 10° of wrist flexion and slight ulnar deviation.

Stability

Palmar Radiocarpal Ligament

From anterior edge of distal radius to proximal carpal row and capitate; checks supinatory movement between joint surfaces and maintains joint integrity (includes the radiocapitate, radiotriquetral, radioscaphoid ligaments).

Palmar Ulnocarpal Ligament

From anterior edge of articular disc and base of the ulnar styloid to carpal—reinforces ulnar side of wrist dorsal radiocarpal ligament; from posterior edge of distal radius to triquetrum and lunate—checks supinatory movement between joint surfaces and maintains joint integrity (includes ulnolunate, ulnotriquetral, ulnocapitate ligaments).

Radial Collateral Ligament

Radial styloid to scaphoid, trapezium, first metacarpal; limits ulnar deviation.

Ulnar Collateral Ligament

Ulnar styloid process to medial triquetrum, pisiform, and articular disc; limits radial deviation.

Dorsal Radiocarpal Ligament

Posterior border of distal radius to scaphoid, lunate, and triquetrum; reinforces dorsal structures.

Fibrous Capsule

Strong but loose; attaches close to articulation and proximally, just distal to inferior epiphyseal line of radius and ulna (figure 9.3, *a*, *b*, and *c*).

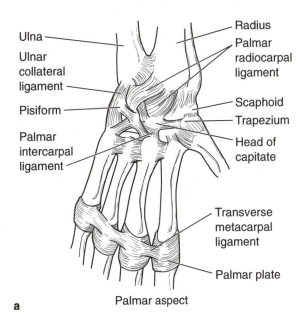

Figure 9.3 (*a*) Palmar view of the wrist and hand ligaments. *(continued)*

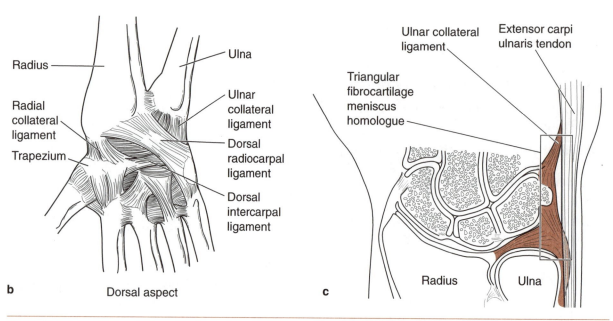

Figure 9.3 (*b*) Dorsal view of the ligaments of the wrist. (*c*) Volar aspect of the triangular fibrocartilage complex.

Special Tests

Phalen's Test

Assessment: Median nerve compression, carpal tunnel syndrome (figure 9.4).

Patient: Sitting.

Test position: Patient places the dorsum of both hands together with the fingers pointing down. Maintain this position for 60 s.

Operator: Note the time of onset of symptoms; test is stopped when symptoms are reproduced.

Positive finding: Pain or paresthesia in the thumb, index finger, and/or other fingers.

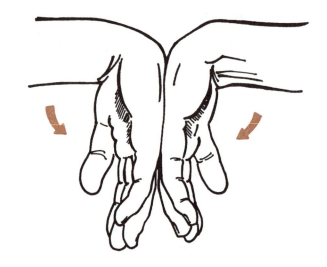

Figure 9.4 Phalen's test.

Median Nerve Tinel's Sign

Assessment: Median nerve (figure 9.5).

Patient: Sitting.

Test position: Forearm in supination.

Operator: Tap over patient's volar carpal ligament with your fingertip.

Positive finding: Pain or paresthesia distal to the wrist.

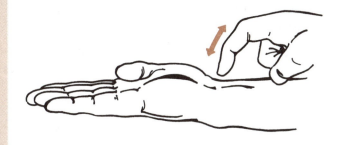

Figure 9.5 Tinel's sign.

Finkelstein's Test

Assessment: Tenosynovitis of the thumb (de Quervain's) (figure 9.6, *a* and *b*).

Patient: Sitting.

Test position: Patient makes a fist with the thumb inside the fingers.

Operator: Stabilize the forearm and ulnarly deviate the wrist.

Positive finding: Pain over the abductor pollicis longus and extensor pollicis brevis.

a

b

Figure 9.6 Finkelstein's test.

Allen Test

Assessment: Patency of the radial and ulnar arteries (figure 9.7).

Patient: Sitting with forearm free to move.

Test position: Elbow bent with fingers pointing up toward the ceiling.

Operator: Compress the radial and ulnar arteries at the wrist, one thumb on the ulnar artery and one thumb on the radial artery. The patient should open and close his or her fist quickly. The operator releases the pressure on one artery and observes the filling pattern of the vessels in the palm. The same is repeated for the other artery.

Positive finding: Blanching remains in the palm after pressure is released from the artery. Can also be used to test individual fingers.

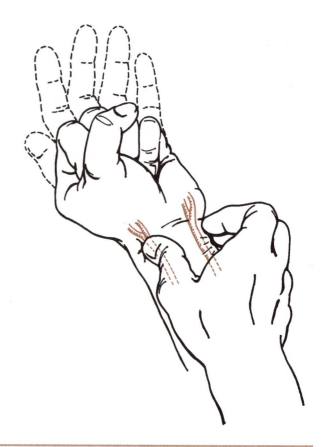

Figure 9.7 Allen test.

MIDCARPAL JOINT

Articulation

Proximally scaphoid, lunate, and triquetrum and distally trapezium, trapezoid, capitate, and hamate.

Type of Joint

Diarthrodial plane joint.

Degrees of Freedom

Primarily glide occurs.

Active Range of Motion

None.

Accessory Movement

Volar Glide

Assessment: To assess extension.

Patient: Sitting with forearm in pronated position.

Operator position: Face the patient, proximal hand on proximal row of carpal bones to be stable, other hand on distal row of carpal bones.

Mobilizing force: Through distal hand in volar direction.

Dorsal Glide

Assessment: To assess flexion.

Patient: Sitting with forearm in supinated position.

Operator position: Face the patient, proximal hand on proximal row of carpal bones to be stable, other hand on distal row of carpal bones.

Mobilizing force: Through distal hand in dorsal direction.

End-Feel

Same as with movements for radiocarpal joints.

Capsular Pattern

None.

Close-Packed Position

Extension with ulnar deviation.

Loose-Packed Position

Neutral or slight flexion with ulnar deviation.

Stability

Palmar Intercarpal Ligament

Connects carpal on palmar side, reinforces palmar arch.

Dorsal Intercarpal Ligament

Connects dorsal proximal and distal row of carpals.

Special Tests

None.

CARPOMETACARPAL JOINT (CMC; 2 TO 5)

Articulation

Distal carpal and proximal metacarpals; second metacarpal with trapezium, trapezoid, capitate; third metacarpal with capitate; fourth with capitate/hamate; fifth with hamate.

Type of Joint

Diarthrodial condyloid joint; fifth CMC is referred to as sellar joint.

Degrees of Freedom

- Flexion and extension.
- Abduction and adduction.

Active Range of Motion

None.

Accessory Movement

Volar/Dorsal Glide

Assessment: To assess volar/dorsal glide of metacarpal for flexion and extension.

Patient: Sitting with forearm supported.

Operator position: Stabilize the carpal with one hand; grasp with the thumb dorsally and index finger volarly. The other hand grasps around the proximal portion of the metacarpal.

Mobilizing force: The thumb on the dorsum of the metacarpal glides the proximal portion of the bone in a volar/dorsal direction.

Distraction

Assessment: To assess mobility (figure 9.8).

Patient: Sitting with forearm supported.

Operator position: Stabilize the carpal with one hand; grasp with the thumb dorsally and index finger volarly. The other hand grasps around the proximal portion of the metacarpal.

Mobilizing force: Apply long-axis distraction to the metacarpal to separate the joint.

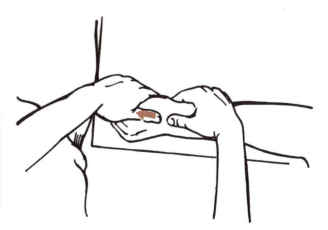

Figure 9.8 Carpometacarpal joint traction.

End-Feel

Not applicable.

Capsular Pattern

Equal limitation in all directions.

Close-Packed Position

Not described.

Loose-Packed Position

Functional position of the wrist.

Stability

Palmar/Dorsal CMC Bands

From the distal carpal bones to metacarpal bases

Interosseous Ligament

From capitate and hamate to the third and fourth metacarpals

Special Tests

None.

ARTHROKINEMATICS OF THE WRIST COMPLEX

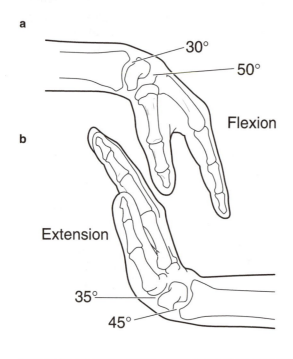

Figure 9.9 (*a*) Flexion of the wrist and (*b*) extension of the wrist.

Extension

45° of the movement occurs at the radiocarpal joint and 25° occurs at the midcarpal joint.

1. Proximal carpal bones move volarly on radius (trapezium and trapezoid bones slide dorsally on scaphoid).
2. Capitate and hamate bones slide volarly on scaphoid, lunate, and triquetrum.
3. Scaphoid spins on the radius at full extension; approximation of scaphoid and lunate to radius and disc.

Flexion

50° of the movement occurs at the midcarpal joint, and 30° occurs at the radiocarpal joint.

1. Proximal carpal bones move dorsally on radius (trapezium and trapezoid move volarly on scaphoid).
2. Capitate and hamate bones slide dorsally on scaphoid, lunate, and triquetrum; distraction of radiocarpal joint.

Ulnar Deviation

Occurs primarily at the radiocarpal joint.

1. Proximal carpal bones glide radially; ulnar glide of hamate and capitate on the triquetrum and lunate.
2. Proximal surface of trapezoid glides radially on the scaphoid; the trapezium and trapezoid move volarly.

Radial Deviation

Occurs primarily at the midcarpal joint.

1. Proximal carpal bones move toward ulna; proximal scaphoid also moves volarly.
2. Proximal surface of capitate and hamate slide ulnarly.
3. Trapezoid moves dorsally.

METACARPAL JOINTS (MCP; 2 TO 5)

Articulation

Convex distal end of each metacarpal and the concave end of each proximal phalanx.

Type of Joint

Diarthrodial condyloid joint.

Degrees of Freedom

- Flexion and extension in sagittal plane through coronal axis that runs through the heads of the metacarpals
- Abduction and adduction in coronal plane through sagittal axis using the third digit as the standard

Active Range of Motion

Flexion: 0-90°

Extension: 0-30°

Abduction: 0-80°

Adduction: 0°

Accessory Movement

Distraction

Assessment: To assess mobility of the joint (figure 9.10).

Patient: Forearm and hand resting on the treatment table.

Operator position: Fixate the proximal bone with the fingers; wrap the fingers and thumb of your other hand around the distal bone close to the joint.

Mobilizing force: Apply long-axis traction to separate the joint surfaces.

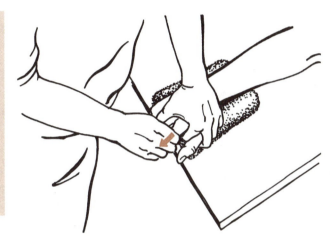

Figure 9.10 Joint traction of a metacarpophalangeal joint.

Glides

Volar glide: To assess flexion (figure 9.11).

Dorsal glide: To assess extension.

Radial glide: To assess abduction.

Ulnar glide: To assess adduction.

Patient: Sitting with forearm supported.

Operator position: Fixate the proximal bone with the fingers; wrap the fingers and thumb of your other hand around the distal bone close to the joint.

Mobilizing force: The glide force is applied by the thumb against the proximal end of the bone.

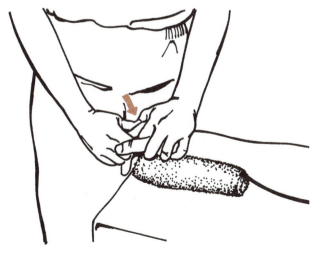

Figure 9.11 Volar glide of a metacarpophalangeal joint.

SECTION

9

Wrist and Hand

End-Feel

Flexion: *Hard* (palmar phalanx contacts metacarpal bones)

Flexion: *Firm* due to tension in dorsal joint capsule and collateral ligaments

Extension: *Firm* due to tension in palmar joint capsule and palmar ligament

Abduction: *Firm* due to tension in collateral ligaments, fascia of web space, and palmar interosseous

Capsular Pattern

Equal restriction in flexion and extension.

Close-Packed Position

Full flexion.

Loose-Packed Position

Slight flexion.

Stability

Fibrous Capsule

Encloses heads of metacarpals and bases of proximal phalanx.

Collateral Ligaments

Run from tubercle and adjacent depression on the side of the metacarpal head to base of phalanx.

Deep Transverse Metacarpal Ligaments

Unite MCP ligaments and hold heads of MCP together.

Volar Plate

Fixed distally to base of phalanx and proximal neck of MCP bone; attaches to deep transverse metacarpal ligament, flexor sheath, and collateral ligaments; prevents hyperextension of the joint.

Special Tests

None.

PROXIMAL INTERPHALANGEAL JOINTS (PIP; 2 TO 5)

Articulation

Convex head of the proximal phalanx and the concave base of the more distal phalanx.

Type of Joint

Diarthrodial hinge joint.

Degrees of Freedom

Flexion and extension in the sagittal plane.

Active Range of Motion

Flexion: 0-120°
Extension: 0-5°

Accessory Movement

Distraction and Glides

Distraction: To assess mobility.

Ventral glide: To assess flexion.

Dorsal glide: To assess extension.

Patient: Sitting with forearm supported.

Operator position: Fixate the proximal bone with the fingers; wrap the fingers and thumb of your other hand around the distal bone close to the joint.

Mobilizing force: The glide force is applied by the thumb against the proximal end of the more distal bone.

End-Feel

Flexion: *Hard* (palmar middle phalanx contacts proximal phalanx)

Flexion: *Soft* (soft tissue contacts soft tissue)

Flexion: *Firm* due to tension in dorsal joint capsule and collateral ligaments

Extension: *Firm* due to tension in palmar joint capsule and palmar ligament

Capsular Pattern

Equal restriction in flexion and extension.

Close-Packed Position

Full extension.

Loose-Packed Position

Slight flexion.

Stability

Collateral Ligaments

Reinforce the sides of the joint.

Special Tests

Volar Plate

Fixed distally to base of phalanx and proximal neck of proximal phalangeal bone, flexor sheath, and collateral ligaments; prevents hyperextension of the joint.

Pinch Test

Assessment: To assess anterior interosseous nerve (figure 9.12).

Patient: Sitting.

Test position: Forearm in pronation and resting on a table.

Operator: The subject is asked to pinch the tips of the index finger and thumb together.

Positive finding: The subject pinches the pulps of the digits instead of the tips.

a

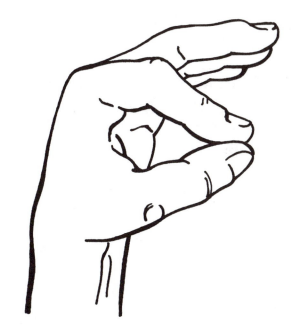

b

Figure 9.12 Pinch test: (*a*) normal pinch and (*b*) abnormal pinch.

Bunnel-Littler Test

Assessment: Tightness or contracture in joint capsule of PIP joint (figure 9.13).

Patient: Sitting.

Test position: Metacarpophalangeal joint held in extension.

Operator: Move the PIP joint into flexion.

Positive finding: The PIP does not move into flexion. If flexed MCP and PIP does move into more flexion, then intrinsic muscle tightens.

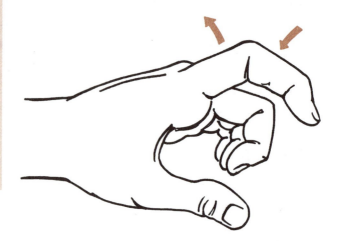

Figure 9.13 Bunnel-Littler test.

Sweater Finger Sign

Assessment: Rupture of the flexor digitorum profundus.

Patient: Sitting.

Test position: Patient is asked to make a fist.

Operator: Assess flexion of the distal phalanx of each finger.

Positive finding: Inability of a finger to flex.

DISTAL INTERPHALANGEAL JOINTS (DIP; 2 TO 5)

Articulation
Convex head of the proximal phalanx and the concave base of the distal phalanx.

Type of Joint
Diarthrodial hinge joint.

Degrees of Freedom
Flexion and extension in the sagittal plane.

Active Range of Motion
Flexion: 0-90°
Extension: 0-10°

Accessory Movement

Distraction and Glides

Distraction: To assess mobility.

Ventral glide: To assess flexion.

Dorsal glide: To assess extension.

Patient: Sitting with forearm supported.

Operator position: Fixate the proximal bone with the fingers; wrap the fingers and thumb of your other hand around the distal bone close to the joint.

Mobilizing force: The glide force is applied by the thumb against the proximal end of the more distal bone.

End-Feel
Flexion: *Firm* due to tension in dorsal joint capsule, collateral ligaments, and oblique retinacular ligament

Extension: *Firm* due to tension in palmar joint capsule and palmar ligaments

Capsular Pattern

Greater limitation in flexion than in extension.

Close-Packed Position

Maximal extension.

Loose-Packed Position

Slight flexion.

Stability

Collateral Ligaments

Reinforce the sides of the joint.

Volar Plate

Fixed distally to base of phalanx and proximal neck of proximal phalangeal bone, flexor sheath, and collateral ligaments; prevents hyperextension of the joint.

Special Tests

None.

ARTHROKINEMATICS OF THE FINGERS

MCP

Flexion

Base of phalanx glides volarly, supination of the phalanx on the metacarpal, digits rotate radially; MCP converge.

Extension

Base of phalanx glides dorsally, reverse of flexion; MCP diverge.

Abduction

Base of phalanx glides toward side that is abducting.

Adduction

Base of phalanx glides toward the side that is adducting.

PIP/DIP

Flexion

Concave base of phalanx glides palmarly.

Extension

Concave base of phalanx glides dorsally.

THUMB CARPOMETACARPAL JOINTS

Articulation

Concavoconvex trapezium (concave: anterior and posterior; convex: medial and lateral) and base of first metacarpal (concave: medial and lateral; convex: anterior and posterior).

Type of Joint

Diarthrodial sellar joint.

Degrees of Freedom

- Flexion and extension in the frontal plane about a dorsal-palmar axis through the trapezium.
- Abduction and adduction in the sagittal plane about a radial ulnar axis through the base of the first CMC.

Active Range of Motion

None measured.

Accessory Movement

Distraction

Assessment: To assess mobility of the joint.

Patient: Forearm and hand resting on the treatment table.

Operator position: Fixate the trapezium with the hand closest to the patient; grasp the patient's metacarpal by wrapping your fingers around it.

Mobilizing force: Apply long-axis traction to separate the joint surfaces.

SECTION 9

Wrist and Hand

118

Glides

Ulnar glide: To assess flexion (figure 9.14).

Radial glide: To assess extension.

Dorsal glide: To assess abduction.

Ventral glide: To assess adduction.

Patient: Sitting with the forearm relaxed.

Operator position: Same as with distraction.

Mobilizing force: The direction of force will depend on the desired glide.

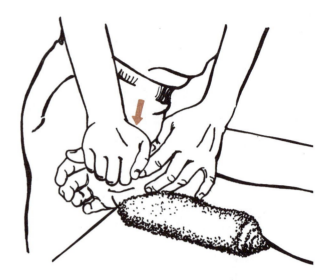

Figure 9.14 Ulnar glide of a carpometacarpal joint of the thumb.

The Clinical Orthopedic Assessment Guide

End-Feel

Flexion: *Soft* due to muscle bulk of the thenar eminence contacting the palm

Flexion: *Firm* due to tension in dorsal joint capsule, short extensors, and short abductors

Extension: *Firm* due to tension in anterior joint capsule and dorsal interosseous

Abduction: *Firm* due to tension in fascia and skin of web and adductor muscle, and dorsal interosseous

Opposition: *Soft* due to muscle bulk of thenar eminence contacting palm

Opposition: *Firm* due to tension in joint capsule, short extensor of thumb, and transverse meta-carpal ligament

Capsular Pattern

Abduction limited most, followed by extension.

Close-Packed Position

Full opposition.

Loose-Packed Position

Midabduction and adduction, and midflexion and extension.

Stability

Anterior and Posterior Oblique Ligaments

From the anterior and posterior surfaces of the trapezium to converge distally to ulnar side of first MCP; anterior is taut in extension and posterior is taut in flexion.

Radial CMC Ligament

From the radial surfaces of the trapezium to the first MCP.

Special Tests

None.

THUMB METACARPAL JOINTS

Articulation

First metacarpal and first proximal phalanx.

Type of Joint

Diarthrodial condyloid joint.

Degrees of Freedom

- Flexion and extension in the frontal plane around an anterior-posterior axis.
- Abduction and adduction in the sagittal plane around a medial-lateral axis.

Active Range of Motion

None measured.

Accessory Movement

Distraction

Assessment: To assess mobility of the joint.

Patient: Forearm and hand resting on the treatment table.

Operator position: Fixate the proximal trapezii complex (trapezium and trapezoid) with the fingers; wrap the fingers and thumb of your other hand around the phalanx of the thumb.

Mobilizing force: Apply long-axis traction to separate the joint surfaces.

Glides

Volar glide: To assess flexion.

Dorsal glide: To assess extension.

Patient: Sitting with forearm supported.

Operator position: Fixate the trapezii complex with the fingers; wrap the fingers and thumb of your other hand around the distal bone close to the joint.

Mobilizing force: The glide force is applied by the thumb against the proximal end of the bone.

End-Feel

Flexion: *Hard* (palmar phalanx is contacting the first metacarpal bone)

Flexion: *Firm* due to tension in dorsal joint capsule, collateral ligament, and short extensor muscles of the thumb

Extension: *Firm* due to tension in palmar joint capsule, palmar ligament, and short flexor muscle of thumb

Capsular Pattern

More limitation in flexion than extension.

Close-Packed Position

Maximal opposition.

Loose-Packed Position

Slight flexion.

Stability

Collateral Ligaments

Reinforce the sides of the joint.

Volar Plate

Fixed distally to base of phalanx and proximal neck of proximal phalangeal bone, flexor sheath, and collateral ligaments; prevents hyperextension of the joint.

Sesamoids

Two sesamoids located on volar surface; maintained by fibers of collateral ligaments and an intersesamoid ligament.

Special Tests

Froment's Sign

Assessment: Adductor pollicis weakness from ulnar nerve paralysis.

Patient: Sitting.

Test position: A piece of paper is placed between the thumb and the radial side of the index finger.

Operator: The operator attempts to remove the paper from the patient's grasp.

Positive finding: The patient is unable to grasp the paper or substitutes with the thumb flexors.

THUMB INTERPHALANGEAL JOINTS (IP)

Articulation

Convex distal end of proximal phalanx and concave proximal end of distal phalanx.

Type of Joint

Diarthrodial hinge joint.

Degrees of Freedom

Flexion and extension.

Active Range of Motion

None.

Accessory Movement

Distraction and Glides

Distraction: To assess mobility.

Ventral glide: To assess flexion.

Dorsal glide: To assess extension.

Patient: Sitting with forearm supported.

Operator position: Fixate the proximal bone with the fingers; wrap the fingers and thumb of your other hand around the distal bone close to the joint.

Mobilizing force: The glide force is applied by the thumb against the proximal end of the bone.

End-Feel

Flexion: *Firm* due to tension in collateral ligaments

Flexion: *Hard* (palmar distal phalanx contacts proximal phalanx)

Extension: *Firm* due to tension in palmar joint capsule and palmar ligament

Capsular Pattern

Greater limitation in flexion than extension.

Close-Packed Position

Maximal extension.

Loose-Packed Position

Slight flexion.

Stability

Collateral Ligaments

Reinforce the sides of the joint.

Volar Plate

Fixed distally to base of phalanx and proximal neck of proximal phalangeal bone, flexor sheath, and collateral ligaments; prevents hyperextension of the joint.

Special Tests

None.

ARTHROKINEMATICS OF THE THUMB

CMC

Flexion

Concave surface of first metacarpal slides ulnarly on convex surface of trapezium; medial rotation of first metacarpal.

Extension

Concave surface of first metacarpal slides radially on convex surface of trapezium; lateral rotation of first metacarpal.

Abduction

Convex surface of first metacarpal slides dorsally on concave surface of trapezium with medial rotation.

Adduction

Convex surface of first metacarpal slides palmarly on concave surface of trapezium with lateral rotation.

MCP

Flexion

Concave base of phalanx glides palmarly.

Extension

Concave base of phalanx glides dorsally.

IP

Flexion

Concave base of phalanx glides palmarly.

Extension

Concave base of phalanx glides dorsally.

Table 9.1 Active Range of Motion of the Wrist and Hand

Joint	Motion	Range of motion
Wrist	Flexion Extension Ulnar deviation Radial deviation	0-80° 0-70° 0-30° 0-20°
MCP (2-5)	Flexion Extension Abduction	0-90° 0-45° 0-30°
PIP (2-5)	Flexion	0-100°
DIP (2-5)	Flexion Extension	0-90° 0-10°
CMC (thumb)	Abduction Flexion	0-70° 0-15°
MCP (thumb)	Flexion	0-50°
DIP (thumb)	Flexion	0-80°

NEUROLOGY

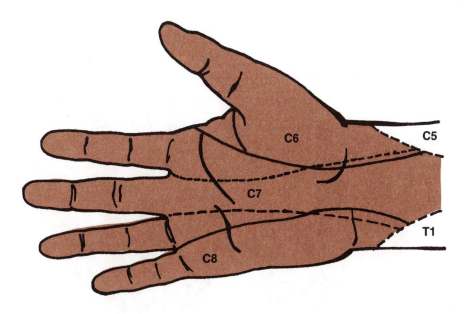

Figure 9.15 Palmar view of hand dermatomes.

Table 9.2 Neurology

Nerve root	Reflex	Muscle	Sensory
C6	Brachioradialis	Wrist extensors	Thumb and index finger
C7	Triceps	Wrist flexors, finger extensors	Middle finger
C8	Abductor digiti minimi	Finger flexors	4th and 5th digits
T1	None	Finger abductors, finger adductors	Medial forearm

Table 9.3 Peripheral Nerves

Nerve	Motor	Sensory
Axillary	Deltoid	Lateral arm, deltoid patch
Median	Thumb pinch, opposition of thumb, thumb abduction	Palmar thumb, 2nd, 3rd, and half of 4th
Radial	Wrist extension, thumb extension	Dorsum of thumb, 2nd, 3rd, and half of 4th
Ulnar	Thumb adduction, little finger abduction	lateral half of 4th and 5th
Musculocutaneous	Biceps	Lateral forearm

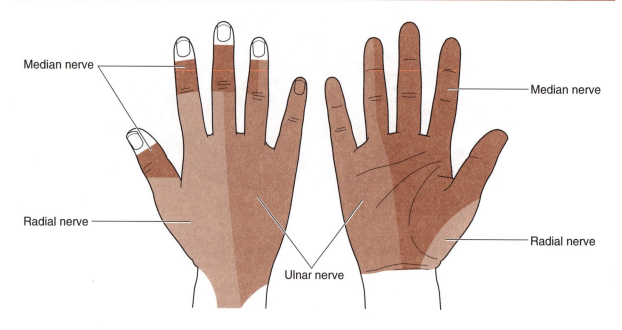

Figure 9.16 Cutaneous nerves of the hand.

The Clinical Orthopedic Assessment Guide

SURFACE PALPATION

Radial styloid process
Anatomical snuff box
Scaphoid
Trapezium
Lister's tubercle
Capitate

Lunate
Ulnar styloid process
Triquetrum
Pisiform
Hook of hamate
Metacarpals
First metacarpal

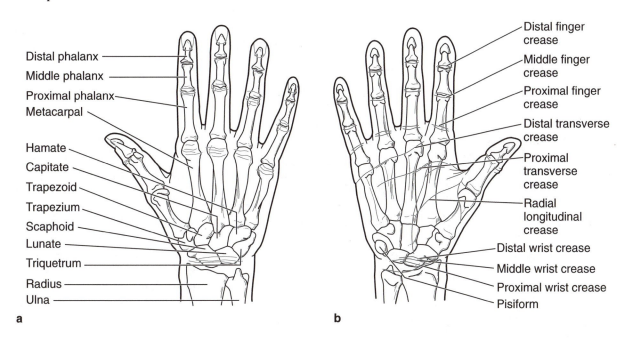

Distal phalanx
Middle phalanx
Proximal phalanx
Metacarpal

Hamate
Capitate
Trapezoid
Trapezium
Scaphoid
Lunate
Triquetrum
Radius
Ulna

a

Distal finger crease
Middle finger crease
Proximal finger crease
Distal transverse crease
Proximal transverse crease
Radial longitudinal crease
Distal wrist crease
Middle wrist crease
Proximal wrist crease
Pisiform

b

Figure 9.17 Boney landmarks and skin creases of the hand: (*a*) dorsal view and (*b*) palmar view.

MUSCLE ORIGIN AND INSERTION

Table 9.4 Muscle Origin and Insertion

Muscle	Origin and insertion
Abductor digiti minimi	Pisiform bone *to* base of proximal phalanx of the little finger
Abductor pollicis brevis	Schapoid, trapezium *to* base of proximal phalanx
Abductor pollicis longus	Middle of radius *to* radial side of base of 1st metacarpal
Adductor pollicis	Capitate, base of 2nd and 3rd metacarpal *to* base of proximal phalanx
Dorsal interossei	Between the metacarpals of the four fingers *to* the base of the proximal phalanx of 2nd, 3rd, and 4th fingers
Extensor carpi radialis brevis	Lateral epicondyle of humerus *to* base of 3rd metacarpal

(continued)

SECTION
9

Wrist and Hand

Table 9.4   *(continued)*

Muscle	Origin and insertion
Extensor carpi radialis longus	Lateral supracondylar ridge of humerus *to* base of 2nd metacarpal
Extensor carpi ulnaris	Lateral epicondyle of humerus *to* base of 5th metacarpal
Extensor digiti minimi	Tendon of extensor digitorum *to* proximal phalanx of little finger
Extensor digitorum	Lateral epicondyle of humerus *to* the dorsal hood of the four fingers
Extensor indicis	Lower ulna and interosseous membrane *to* dorsal hood of index finger
Extensor pollicis brevis	Middle of radius and ulna *to* base of proximal phalanx of thumb
Extensor pollicis longus	Middle third of ulna and interosseous membrane *to* base of distal phalanx of thumb
Flexor carpi radialis	Medial epicondyle of humerus *to* base of 2nd and 3rd metacarpals
Flexor carpi ulnaris	Medial epicondyle *to* pisiform, hamate, base of 5th metacarpal
Flexor digiti minimi	Hamate bone *to* proximal phalanx of little finger
Flexor digitorum profundus	Anterior, medial ulna *to* base of distal phalanx of four fingers
Flexor digitorum superficialis	Medial epicondyle *to* base of middle phalanx of four fingers
Flexor pollicis brevis	Trapezium, trapezoid, capitate *to* base of proximal phalanx of thumb
Flexor pollicis longus	Middle radius and interosseous membrane *to* base of distal phalanx of thumb
Lumbricales	Tendon of flexor digitorum profundus *to* dorsal hood of the four fingers
Opponens digiti minimi	Hamate bone *to* 5th metacarpal
Opponens pollicis	Trapezium *to* 1st metacarpal
Palmar interrossei	Sides of 2nd, 4th, and 5th metacarpals *to* base of proximal phalanx of same fingers
Plamaris longus	Medial epicondyle *to* palmar aponeurosis

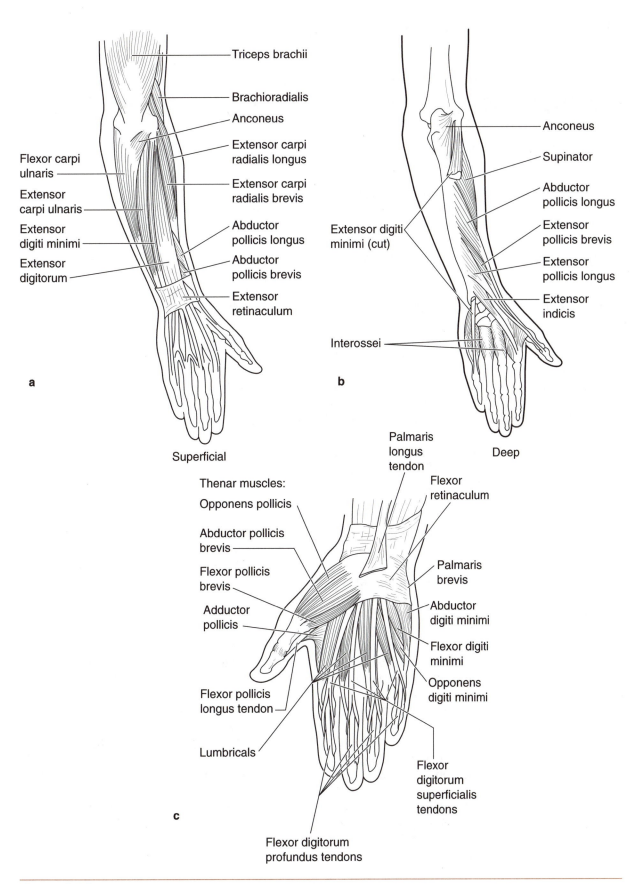

Triceps brachii

Brachioradialis

Anconeus

Extensor carpi radialis longus

Extensor carpi radialis brevis

Abductor pollicis longus

Abductor pollicis brevis

Extensor retinaculum

Flexor carpi ulnaris

Extensor carpi ulnaris

Extensor digiti minimi

Extensor digitorum

a

Superficial

Anconeus

Supinator

Abductor pollicis longus

Extensor pollicis brevis

Extensor pollicis longus

Extensor indicis

Extensor digiti minimi (cut)

Interossei

b

Deep

Thenar muscles:

Opponens pollicis

Abductor pollicis brevis

Flexor pollicis brevis

Adductor pollicis

Flexor pollicis longus tendon

Lumbricals

Palmaris longus tendon

Flexor retinaculum

Palmaris brevis

Abductor digiti minimi

Flexor digiti minimi

Opponens digiti minimi

Flexor digitorum superficialis tendons

Flexor digitorum profundus tendons

c

Figure 9.18 Muscles acting on the wrist. (*a*) Superficial view and (*b*) deep views of the muscles of the lower arm and wrist. (*c*) Deep view of the muscles of the hand.

MUSCLE ACTION AND INNERVATION

Table 9.5   Muscle Action and Innervation

Action	Muscle involved	Nerve supply	Nerve root
Wrist flexion	Flexor carpi radialis Flexor carpi ulnaris	Median Ulnar	C6-C7 C7-C8
Wrist extension	Extensor carpi radialis longus Extensor carpi radialis brevis Extensor carpi ulnaris	Radial Radial Radial	C6-C7 C7-C8 C7-C8
Ulnar deviation	Flexor carpi ulnaris Extensor carpi ulnaris	Ulnar Radial	C7-C8 C7-C8
Radial deviation	Flexor carpi radialis Extensor carpi radialis longus Abductor pollicis longus Extensor pollicis brevis	Median Radial Radial Radial	C6-C7 C6-C7 C7-C8 C7-C8
Finger extension	Extensor digitorum communis Extensor indices (2nd digit) Extensor digiti minimi (5th digit)	Radial Radial Radial	C7-C8 C7-C8 C7-C8
Finger flexion	Flexor digitorum profundus Flexor digitorum superficialis Lumbricals Interossei Flexor digiti minimi (5th digit)	Median Median 1-2 median 3-4 ulnar Ulnar Ulnar	C8, T1 C7-C8, T1 C8, T1 C8, T1 C8, T1
Finger abduction	Dorsal interossei Abductor digiti minimi	Ulnar Ulnar	C8, T1 C8, T1
Finger adduction	Palmar interossei	Ulnar	C8, T1
Thumb extension	Extensor pollicis longus Extensor pollicis brevis Abductor pollicis longus	Radial Radial Radial	C7-C8 C7-C8 C7-C8
Thumb flexion	Flexor pollicis brevis Flexor pollicis longus Opponens pollicis	Superficial head: median Deep head: ulnar Median Median	C8, Ti C8, T1 C8, T1 C8, T1
Thumb abduction	Abductor pollicis longus Abductor pollicis brevis	Radial Median	C7-C8 C8, T1

The Clinical Orthopedic Assessment Guide

Table 9.5 *(continued)*

Action	Muscle involved	Nerve supply	Nerve root
Thumb adduction	Adductor pollicis	Ulnar	C8, T1
Opposition	Opponens pollicis Flexor pollicis brevis Abductor pollicis brevis Opponens digiti minimi	Median Median Median Ulnar	C8, T1 C8, T1 C8, T1 C8, T1

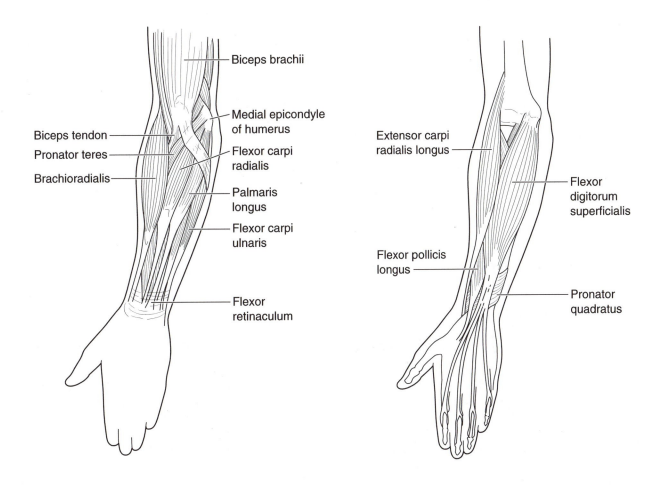

a　　　　　　　　　　　　　　　　b

Figure 9.19 Ventral view of muscles acting on the wrist and fingers: (*a*) superficial view; (*b*) intermediate view.

(continued)

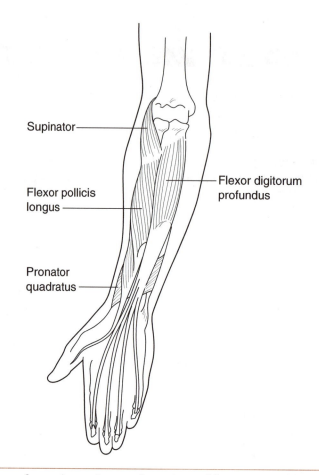

c

Figure 9.19 Ventral view of muscles acting on the wrist and fingers: (c) deep view.

DIFFERENTIAL DIAGNOSIS

Table 9.6 Differential Diagnosis

Disorder	Description	Onset	Symptoms	Signs	Special tests
de Quervain's syndrome	Tendonitis of the abductor pollicis longus and exterior pollicis brevis	Overuse of these tendons in activities requiring repeated radial deviation	Pain over the APL and EPB	Pain to palpation; swelling; crepitus in tendon region	Finkelstein's test
Carpal tunnel syndrome	Constriction of the carpal tunnel that houses the median nerve	Overuse of wrist with activities such as wrist flexion; trauma	Pain numbness, nocturnal pain along the median nerve distribution	Possible atrophy of hand muscles innervated by the median nerve	Tinel's test; Phalen's test

Table 9.6 *(continued)*

Disorder	Description	Onset	Symptoms	Signs	Special tests
Bennett's fracture	Fracture of 1st metacarpal	Direct force from a punch	Pain along MCP shaft	Swelling, deformity, tender to palpation	
Boxer's fracture	Fracture of the 5th metacarpal	Direct force from a punch	Pain	Swelling, deformity	
Mallet finger	Extensor tendon avulsed from DIP	Direct force causing forced flexion of distal phalanx	Pain	Deformity in which the DIP cannot straighten	Sweater finger test
Colles fracture	Fracture of distal radius may or may not include ulna with or without displacement	Direct trauma	Pain; watch for impaired circulation	Dinner-fork deformity; swelling	
Kienboch's disease	Avascular necrosis of the lunate	Idiopathic, may be repeated minor trauma	Wrist pain	Synovitis	Bone scan
Scaphoid fracture	Fracture of scaphoid	Direct force causing hyper-extension of the wrist	Pain in the anatomical snuff box	Tender to palpation, swelling	Point tenderness in anatomical snuff box
Raynaud's phenomenon	Constriction of the distal vessels	Trauma, idiopathic	Decreased circulation distally; pain	Blanching of fingers	
Reflex sympathetic dystrophy	Increased sympathetic output	Following trauma	Constant pain and burning	Joint swelling, stiffness, discoloration (red or blue), hyperhydrosis	
Skier's thumb, gameskeeper's thumb	Sprain of the ulnar collateral	Direct force causing hyperexten-	Pain at base 1st metacarpal	Possible deformity, increased	

(continued)

Table 9.6 *(continued)*

Disorder	Description	Onset	Symptoms	Signs	Special tests
Skier's thumb; gameskeeper's thumb *(continued)*	ligament	sion and abduction of the 1st MCP joint, may also be repeated trauma		laxity in 1st MCP joint	
Ulnar neuropathy	Prolonged pressure on the ulnar nerve as it passes around the hook of the hamate	Direct pressure from prolonged wrist extension, such as with bike riding or repeated trauma to ulna	Tingling, paresthesia of the hands and fingers along ulnar distribution	Possible atrophy of hypothenar eminence, interosseous, medial 2 lumbricales, thumb adduction	Froment's sign
Smith's fracture	Fracture of the distal end of radius with volar displacement	Direct force	Pain	Possible deformity, tender to palpation	
Osteoarthritis	Degenerative changes in synovial joint	Overuse	Pain, stiffness in joints	Deformity, swelling	X-rays
Dupuytren's contracture	Palmar fascia constriction	Idiopathic, insidious	Inability to straighten fingers, primarily 4th and 5th	Nodules in palmar fascia	

SECTION 10

FUNCTIONAL ASSESSMENT OF THE UPPER EXTREMITY

The present emphasis in health care is on functional recovery of the patient. It is thus important to use valid and reliable testing tools to report progress in therapy. This section presents functional assessment tools that have been extensively used by orthopedic practitioners. Functional ranges needed for activities of daily living (ADLs) are listed for the shoulder, elbow, hand, and wrist. Last in this section is a chart on throwing mechanics that should be helpful for practitioners who deal with athletes who throw.

SHOULDER

Functional Ranges

Reaching overhead: shoulder flexion to 160°

Reaching behind back (reaching into back pocket): full internal rotation

Placing hand behind head (combing hair): full external rotation

Wash opposite axilla: full horizontal adduction

Functional Strength Testing of the Shoulder

Lateral Scapular Slide Test (Kibler, Davies)

Assessment: Quantify bilateral comparison of the scapula.

Patient: Standing.

Test position: Five test positions:

1. Arm at side
2. Hands on hip with thumb posterior
3. Empty can
4. 120° of elevation in scapular plane
5. 150° of elevation in scapular plane

Operator: Measure the distance from T7 to inferior angle of the scapula in five positions.

Positive finding: Greater than 1 cm of asymmetry in a bilateral comparison.

Table 10.1 Functional Tests

Starting position	Action	Functional test*
Sitting	Forward flex arm to 90°	Lift 4 to 5 lb: functional Lift 1 to 3 lb: functionally fair Lift arm weight: functionally poor Cannot lift arm: nonfunctional
Sitting	Shoulder extension	Lift 4 to 5 lb: functional Lift 1 to 3 lb: functionally fair Lift arm weight: functionally poor Cannot lift arm: nonfunctional
Side-lying	Shoulder internal rotation (can sit and use pulley)	Lift 4 to 5 lb: functional Lift 1 to 3 lb: functionally fair Lift arm weight: functionally poor Cannot lift arm: nonfunctional
Side-lying	Shoulder external rotation (can sit and use pulley)	Lift 4 to 5 lb: functional Lift 1 to 3 lb: functionally fair Lift arm weight: functionally poor Cannot lift arm: nonfunctional
Sitting	Shoulder abduction	Lift 4 to 5 lb: functional Lift 1 to 3 lb: functionally fair Lift arm weight: functionally poor Cannot lift arm: nonfunctional
Sitting	Shouder abduction with wall pulley	Lift 4 to 5 lb: functional Lift 1 to 3 lb: functionally fair Lift arm weight: functionally poor Cannot lift arm: nonfunctional
Sitting	Shoulder elevation (shoulder shrug)	Lift 4 to 5 lb: functional Lift 1 to 3 lb: functionally fair Lift arm weight: functionally poor Cannot lift arm: nonfunctional
Sitting	Sitting push-up (shoulder depression)	Lift 4 to 5 lb: functional Lift 1 to 3 lb: functionally fair Lift arm weight: functionally poor Cannot lift arm: nonfunctional

* Younger patients may be able to lift more weight than listed. Use opposite side for accurate comparison.

The Clinical Orthopedic Assessment Guide

Davies Functional Shoulder Rating Scale

Symptoms (Analogue Pain Scale)   10

Pain at rest (0-10) ——
Pain with activity (0-10) ——

Average score at rest; with activity, reverse numbers for actual score on scale (i.e., 3 = 7).

Patient's Functional Rating   20

Return to former sport with no limitations or pain = 20

Return to former sport with no limitation but pain = 17-19

Return to former sport with limitation but no pain = 14-16

Return to former sport with limitation and pain = 11-13

Inability to return to former sport level but still plays sport = 8-10

Ability to return to less competitive sports with no pain = 4-7

Inability to return to any sport = 0

Physical Exam   10

Each of these tests can be found in the guide to the shoulder (section 7). Give one point for each positive test; if the finding is significant, add an additional point; reverse numbers for actual score on scale.

Biceps brachii test ——
Rotator cuff tests ——
Impingement tests ——
Anterior instability tests ——
Posterior instability tests ——
Multi-directional instability tests ——

Active Range of Motion (ROM)   20
(Within normal limits: bilateral comparison)

Flexion (0-180°) ——
Abduction (0-180°) ——
Internal rotation (0-70°) ——

External rotation (0-90°) ——
Scaption (0-90°) ——
All motions totaled together: ——

0% diff = 20
< 5% = 18
< 10% = 16
< 15% = 14
< 20% = 12
< 25% = 10
< 30% = 8
> 30% = 1-7

Isokinetic Testing   20

Internal rotation ——
External rotation ——
Agonist/antagonist ratio ——

All tests totaled together—peak torque, total work, average power: ——

0% diff = 20
< 5% = 18
< 10% = 16
< 15% = 14
< 20% = 12
< 25% = 10
< 30% = 8
> 30% = 1-7

Kinesthetic Testing   10

Flexion < 90° ——
Flexion > 90° ——
Abduction < 90° ——
Abduction > 90° ——
Internal rotation < 45° ——
External rotation < 45° ——
External rotation > 45° ——

Absolute number comparison in degrees, all tests totaled together and divided by seven tests, ability to reproduce specified angle with eyes closed: ——

< 5 = 10
< 7 = 9
< 9 = 8

(continued)

SECTION 10

Upper Extremity

```
< 11 = 7
< 13 = 6
< 15 = 5
< 17 = 4
< 19 = 3
< 21 = 2
> 21 = 1
```

Functional Throwing Performance Index (FTPI)   *10*

Test: A 1 ft x 1 ft square is displayed on a wall, 4 ft up from the floor; individual stands 15 ft from square and throws a 20 in. circumference rubber playground ball.

Protocol:

1. Use crow-hop technique.
2. Perform four warm-ups.
3. Patient performs five maximally con trolled throws, and must also catch the ball on the rebound.
4. The patient throws as many times as she or he can in 30 s.
5. Three 30 s tests are performed.
6. Analysis: Number of throws and number within square are recorded.

FTPI = accuracy in target   ____

Total number of throws x 100   ____

TOTAL (out of 100 points)   ____

Davies Functional Shoulder Rating Scale

Rating	Point range
Excellent	80-100
Good	70-84
Fair	55-69
Poor	< 55

Functional Assessment of the Shoulder

Instructions: For each portion of this assessment, circle the appropriate number of points. Add the number of points scored after each test and divide by the total points available.

Pain	Points (select one)
None	15
Mild	10
Moderate	5
Severe	0
(Total points available)	**15**
Total scored	____

ADLs Activity Level	Points (select all that apply)
Full work	4
Full recreation or sport	4
Unaffected sleep	2
Positioning	
Up to waist	2
Up to xiphoid	4
Up to neck	6
Up to top of head	8
Above head	10
(Total points available)	**20**
Total scored	____

Range of Motion Abduction	Points (select one)
0-30°	0
31-60°	2
61-90°	4
91-120°	6
121-150°	8
151-180°	10
Flexion	**Points** (select one)
0-30°	0
31-60°	2
61-90°	4
91-120°	6
121-150°	8
151-180°	10
Internal Rotation	**Points** (select one)
Dorsum of hand to lateral thigh	0

Dorsum of hand to buttock	2
Dorsum of hand to lumbosacral junction	4
Dorsum of hand to waist (L3)	6
Dorsum of hand to T12	8
Dorsum of hand to T7	10

External Rotation **Points** (select all that apply)	
Hand behind head with elbow held forward	2
Hand behind head with elbow held back	2
Hand on top of head with elbow held forward	2

Hand on top of head with elbow held back	2
Full elevation from on top of head	2
(Total points available)	**40**
Total scored	—
Power	**25 lb**
Amount of weight lifted	—

Reprinted, by permission, from C.R. Constant and A.H.G. Murley, 1987, "A clinical method of functional assessment of the shoulder," *Clinical Orthopaedics Related Research* 214: 160-164.

ELBOW

Functional Range of Motion

Table 10.2 Functional Range of Motion

Activity	Flexion	Pronation	Supination
Opening a door	30-60°	0-35°	0-25°
Putting on shoes	10-20°	10-30°	None needed
Pouring pitcher	35-55°	0-40°	0-25°
To reach sacrum	65-75°	None needed	45-75°
Lifting a chair	20-95°	10-35°	None needed
To reach waist	90-105°	None needed	0-30°
To hold a newspaper	75-100°	10-50°	None needed
To reach top of head	110-125°	None needed	40-60°
To operate a knife for eating	90-110°	30-40°	None needed
To operate a fork for eating	80-125°	0-15°	0-50°
To drink from a glass	40-130°	0-10°	0-10°
To reach occiput of head	135-150°	0-15°	0-15°
To use a telephone	40-135°	0-40°	0-25°

SECTION **10**

Upper Extremity

Functional Assessment of the Elbow

Table 10.3   Functional Tests

Starting position	Action	Functional test
Sitting	Bring hand to mouth lifting weight (elbow flexion)	Lift 5 to 6 lb: functional Lift 3 to 4 lb: functionally fair Lift 1 to 2 lb: functionally poor Lift 0 lb: nonfunctional
Standing 3 ft from wall, lean against wall	Push arms straight (elbow extension)	5 to 6 reps: functional 3 to 4 reps: functionally fair 1 to 2 reps: functionally poor 0 reps: nonfunctional
Stand, facing closed door	Open door starting with palm down (supination)	5 to 6 reps: functional 3 to 4 reps: functionally fair 1 to 2 reps: functionally poor 0 reps: nonfunctional
Stand, facing closed door	Open door starting with palm up (pronation)	5 to 6 reps: functional 3 to 4 reps: functionally fair 1 to 2 reps: functionally poor 0 reps: nonfunctional

Adapted from Palmer ML, and Epler M. Clinical Assessment Procedures in Physical Therapy. Philadelphia, JB Lippincott, 1990

WRIST, HAND, AND GRIP

Functional Ranges

The amount of motion needed to perform most activities with the hand is 10° of wrist extension and 30° of wrist flexion. Figure 10.1 illustrates different types of grips.

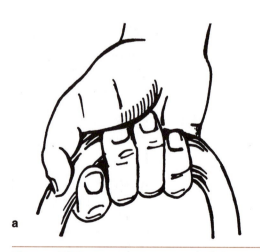
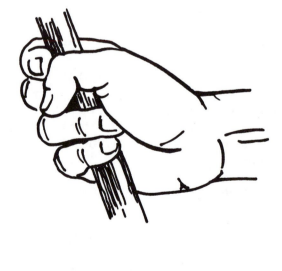

a   b

Figure 10.1   Fine motor control of the hand: (*a*) hook grasp, (*b*) fist grasp.

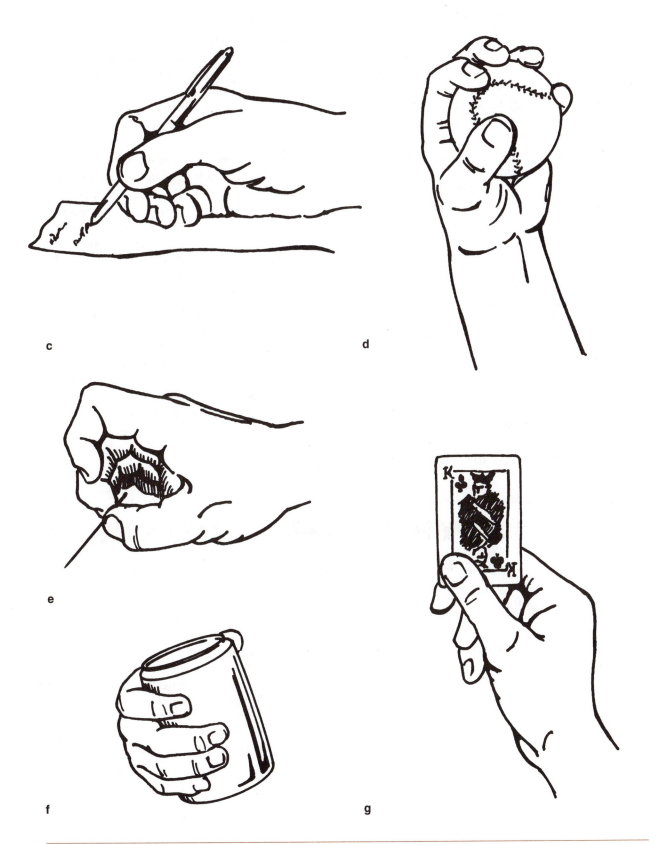

c

d

e

f

g

Figure 10.1 Fine motor control of the hand: (*c*) palmar prehension, (*d*) spherical grasp, (*e*) tip prehension, (*f*) cylindrical grasp, and (*g*) lateral prehension.

Nonprehensile: Use of hand in a nongrasp capacity; contact is with palmar portion of hand(s).

Prehension: Requires grasping or taking hold of an object between two surfaces of the hand; the thumb may or may not be involved.

Power grip: A grip that is used to hold onto an object but is moved with proximal joints.

a. *Hook grip:* The fingers are used as a hook as in carrying a purse or briefcase (MCPs neutral, flexion at PIP and DIP, thumb extended, use of flexor digitorum profundis [FDP] and flexor digitorum superficialis [FDS])

b. *Cylindrical grip:* The entire palmar surface surrounds a cylindrical object such as a glass. The thumb is involved (fingers adducted and flexed, thumb opposed, use of flexor pollicis longus [FPL], adductor pollicis, fourth lumbrical, FDP).

c. *Spherical grip:* The grasp is adjusted to a spherical object such as a ball (fingers adducted and flexed, thumb opposed, use of finger flexors).

d. *Lateral prehension:* A grip that requires two adjacent fingers to adduct to enclose the object as in holding a newspaper (thumb and index finger adducted, lumbricals).

Precision: Grips that require skillful placement of fingers in order to manipulate an object.

a. *Two-point tip pinch:* Tip-to-tip grip, involving thumb and one other finger as in picking up a needle (thumb opposed and flexed; fingers flexed at MCP, PIP, DIP; use of FDP, FPL, interossei).

b. *Three-point tip pinch (three-jaw chuck):* The tip of the thumb is used in conjunction with digits 2 and 3 (similar to two-point tip pinch).

c. *Two-point pad pinch:* The thumb and the pad of the index finger are used to maneuver an object such as a light bulb (thumb opposed and flexed, fingers flexed at MCP and PIP, flexed or extended at DIP).

d. *Three-point pad pinch (three-jaw chuck):* The thumb opposes digits 2 and 3 with the pads of the distal phalanges as in unscrewing a bottle top (similar to two-point pad pinch).

e. *Lateral pinch (lateral prehension):* A thin object is grasped between the thumb and the lateral side of the index finger as when using a key (thumb adducted with IP flexion, index finger flexed and abducted).

Functional Assessment of the Wrist and Hand

Table 10.4   Functional Tests

Starting position	Action	Functional test
Forearm supinated, resting on table	Wrist flexion	Lift 5+ lb: functional Lift 3 to 4 lb: functionally fair Lift 1 to 2 lb: functionally poor Lift 0 lb: nonfunctional
Froearm pronated, resting on table	Wrist extension lifting 1 to 2 lb	5+ reps: functional 3 to 4 reps: functionally fair 1 to 2 reps: functionally poor 0 reps: nonfunctional
Forearm neutral, resting on table	Radial deviation lifting 1 to 2 lb	5+ reps: functional 3 to 4 reps: functionally fair 1 to 2 reps: functionally poor 0 reps: nonfunctional

Table 10.4 *(continued)*

Starting position	Action	Functional test
	Thumb flexion with resistance from 1 cm wide rubber band	5+ reps: functional 3 to 4 reps: functionally fair 1 to 2 reps: functionally poor 0 reps: nonfunctional
Forearm neutral, rubber band around thumb and index finger	Thumb extension against rubber band	5+ reps: functional 3 to 4 reps: functionally fair 1 to 2 reps: functionally poor 0 reps: nonfunctional
Forearm resting on table, rubber band around thumb and index finger	Thumb abduction against rubber band	5+ reps: functional 3 to 4 reps: functionally fair 1 to 2 reps: functionally poor 0 reps: nonfunctional
Forearm resting on table	Thumb adduction, lateral pinch of piece of paper	Hold 5 s: functional Hold 3 to 4 s: functionally fair Hold 1 to 2 s: functionally poor Hold 0 s: nonfunctional
Forearm resting on table	Thumb opposition, pad pinch of piece of paper	Hold 5 s: functional Hold 3 to 4 s: functionally fair Hold 1 to 2 s: functionally poor Hold 0 s: nonfunctional
Forearm resting on table	Finger flexion, grasping mug with cylindrical grip, lifting off table	5+ reps: functional 3 to 4 reps: functionally fair 1 to 2 reps: functionally poor 0 reps: nonfunctional
Forearm resting on table	Put on rubber gloves, keeping fingers straight	2 to 4 s: functional 4 to 8 s: functionally fair 10 to 20 s: functionally poor 21+ s: nonfunctional
Forearm resting on table	Finger abduction against resistance of rubber band	Hold 5+: functional Hold 3 to 4: functionally fair Hold 1 to 2: functionally poor Hold 0: nonfunctional
Forearm resting on table	Finger adduction of piece of paper while examiner tries to pull it out	Hold 5+: functional Hold 3 to 4: functionally fair Hold1 to 2: functionally poor Hold 0: nonfunctional

SECTION
10

Upper Extremity

139

THROWING MECHANICS

Table 10.5 Throwing Mechanics

Shoulder and arm movements	Shoulder muscles involved	Sructures under stress
Wind-up: Phase that prepares the pitcher for correct body posture and balance. The body rotates away from the plate.	Minimal shoulder activity.	There are minimal acceleration and deceleration forces, therefore minimal injury in this phase.
Cocking: The contralateral leg is placed forward, the pelvis is internally rotated, and chest is forward. Shoulder abducted to 90°, humerus externally rotated 100° to 120°, elbow at 90°, wrist extended.	Posterior deltoid horizontally abducts the humerus. Supraspinatus, infraspinatus, and teres minor stabilize humeral head.	Tension is developed in anterior capsule, anterior deltoid, pectoralis insertion. The long head of the biceps is also under tension.
Acceleration: The body is brought forward with arm trailing. The ball is accelerated forward. Shoulder moves through internal rotation. Wrist is snapped from extended position.	The subscapularis, pectoralis major, latissimus dorsi, and teres major contract concentrically. The supraspinatus, infraspinatus, and teres minor work eccentrically to control horizontal adduction. Elbow flexors work eccentrically to control elbow extension.	Walgus stress is placed on elbow. Rotational stress is placed on the humeral shaft. Impingement of supraspinatus tendon and/or other subacromial soft tissue structures. Forearm flexors pull on the medial epicondyle.
Release/deceleration: Deceleration occurs following ball release to maintain the integrity of the shoulder. Initially, the humerus is internally rotating and the elbow is extending.	Subscapularis, pectoralis major, latissimus dorsi, and triceps. Pronator teres, forearm flexors or supinators to control forearm momentum.	Posterior capsule and posterior deltoid under stress. Possible subluxation of glenohumeral joint. The long head of the biceps may be stressed with certain pitches (curve ball).
Follow-through: The body moves forward and the ipsilateral foot is planted ahead of the body.	The posterior deltoid and posterior rotator cuff contract eccentrically to slow down arm.	No significant shoulder injuries.

PART IV

LOWER EXTREMITY

THE HIP JOINT

The hip joint is similar to the shoulder joint in anatomy, but the hip is much more stable because of its strong ligament support and deeper articulating fossa. The hip articulation is primarily weight bearing and can withstand great compressive loads.

JOINT BASICS

Articulation

Convex head of the femur and concave acetabulum of the pelvis.

Type of Joint

Diarthrodial spheroidal joint.

Degrees of Freedom

- Flexion and extension in sagittal plane about a coronal axis through the femoral head.
- Abduction and adduction in coronal plane about a sagittal axis through the femoral head.
- Internal rotation and external rotation in transverse plane about a longitudinal axis through the femoral head.

Active Range of Motion (AAOS)

Hip flexion: 0-120°

Hip extension: 0-30°

Hip abduction: 0-45°

Hip adduction: 0-30°

Hip internal rotation: 0-45°

Hip external rotation: 0-45°

Accessory Movement

Caudal glide

Assessment: To assess joint mobility.

Patient: The patient is supine with the hip and knee flexed and supported by the practitioner's shoulder.

Operator position: Wrap your hands around the proximal thigh.

Mobilizing force: A caudal mobilizing force on the proximal femur is performed as the practitioner leans back. Caudal glide increases the range of motion of the hip and decreases joint pain.

Posterior Glide

Assessment: To assess flexion and internal rotation of the hip (figure 11.1).

Patient: The patient is supine with the hip in the resting position (30° flexion, 30° abduction, slight external rotation).

Operator position: The mobilizing hand contacts the anterior aspect of the proximal femur.

Mobilizing force: Apply the mobilizing force straight down in a posterior direction. Posterior glide is necessary for flexion and internal rotation.

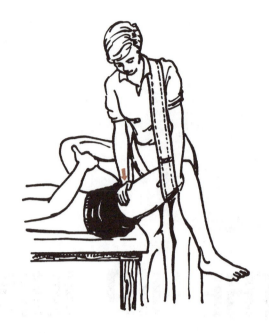

Figure 11.1 Posterior glide of the hip joint.

Anterior Glide

Assessment: To assess extension and external rotation (figure 11.2).

Patient: The patient is prone with the hip off the edge of the table.

Operator position: Hold the leg with the distal hand. The proximal hand is placed posteriorly, just inferior to the buttock.

Mobilizing force: The proximal hand applies an anterior force. Anterior glide is necessary for extension and external rotation.

Figure 11.2 Anterior glide of the hip joint.

Lateral Glide

Assessment: To assess lateral mobility of the hip (figure 11.3).

Patient: Supine with leg extended (can be flexed to 90°).

Operator position: One hand stabilizes the lateral aspect of the distal femur and the other hand contacts the medial aspect of the proximal femur.

Mobilizing force: The proximal hand applies a lateral force.

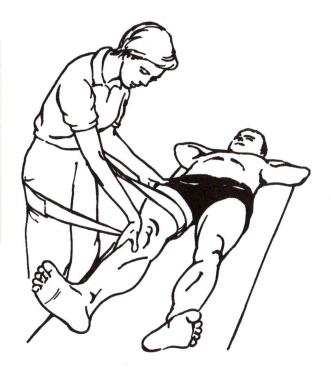

Figure 11.3 Compression of the hip joint.

End-Feel

Flexion: *Soft* (tissue approximation) due to muscle bulk of anterior thigh and lower abdominals or stretch of hip extensors.

Extension: *Firm* due to issue stretch of anterior capsule, iliofemoral, ischiofemoral and pubofemoral ligaments, and hip flexor muscles.

Abduction: *Firm* due to tissue stretch of inferior capsule, pubofemoral and ischiofemoral ligaments, inferior band of iliofemoral ligament, and adductor muscles.

Abduction: *Hard* (femoral neck approximates the acetabulum).

Adduction: *Soft* (tissue approximation of thighs).

Adduction: *Firm* due to stretch of abductor muscles.

Internal rotation: *Firm* due to tissue stretch of posterior joint capsule, ischiofemoral ligament, and external rotators of the hip.

External rotation: *Firm* due to tissue stretch of anterior joint capsule, iliofemoral and pubofemoral ligaments, and internal rotator muscles.

Capsular Pattern

Equal limitation in flexion, abduction, and internal rotation with slight loss in extension, little or no loss in external rotation.

Close-Packed Position

Maximum extension, internal rotation, slight abduction.

Loose-Packed Position

30° of flexion, 30° of abduction, slight external rotation.

Stability

Iliofemoral Ligament (Y Ligament)

From anterior inferior iliac spine down to the upper and lower parts of the intertrochanteric line of the femur; blends with the anterior capsule, prevents overextension during standing and external rotation; during adduction the medial inferior band is under mild tension.

Ischiofemoral Ligament

From the body of the ischium near the acetabular margin upward and laterally to the greater trochanter; limits extension, internal rotation, and abduction.

Pubofemoral Ligament

From the superior ramus of the pubis down to the lower part of the intertrochanteric line; limits extension, abduction, and external rotation (figure 11.4, *a* through *e*).

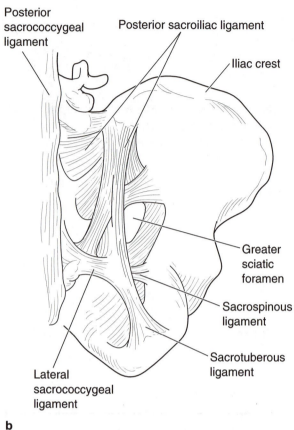

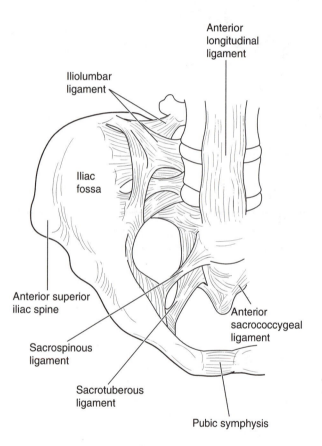

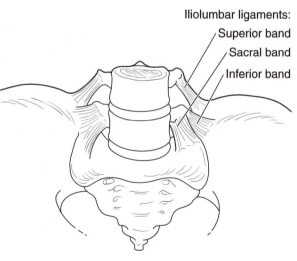

Figure 11.4 Pelvis ligaments: (*a*) anterior view and (*b*) posterior view. (*c*) Iliolumbar ligaments.

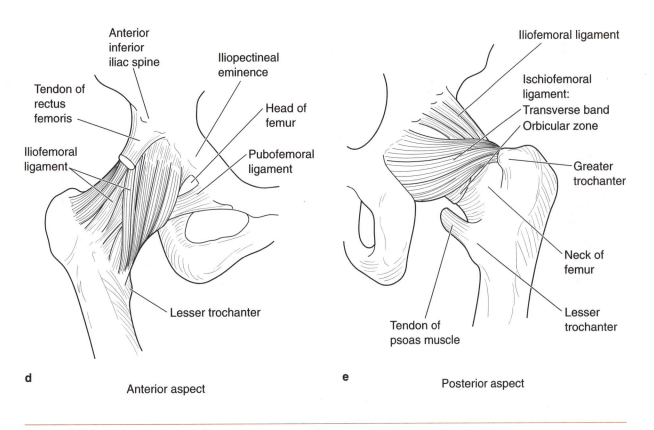

Figure 11.4 Hip ligaments: (*d*) anterior view and (*e*) posterior view.

Special Tests

Patrick's or Faber Test

Assessment: Range of motion of hip and pain (figure 11.5).

Patient: Supine with leg placed so foot is resting on top of knee of opposite leg.

Test position: Hip in position of flexion, abduction, external rotation.

Operator: Slowly lower test leg in abduction toward table.

Positive finding: Leg does not reach table or pain is present in this position.

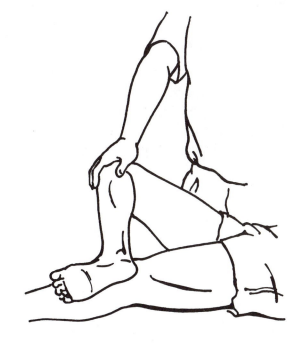

Figure 11.5 Anterior glide of the hip joint.

Craig's Test

Assessment: Measures femoral anteversion (figure 11.6).

Patient: Prone.

Test position: Knee flexed to 90°.

Operator: Passively move limb from internal to external rotation of the hip, while palpating greater trochanter. When the trochanter reaches its most lateral position, measure the angle between the lower leg and vertical.

Positive finding: In the adult the mean angle is 8° to 15°.

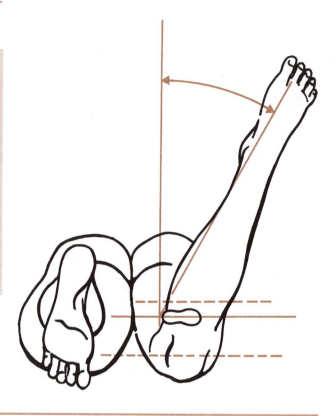

Figure 11.6   Craig's test.

Quadrant or Scour Test

Assessment: Pathology in articulating surface of the hip joint (figure 11.7).

Patient: Supine.

Test position: Hip in flexion with knee bent.

Operator: Apply slight resistance while hip is taken into an arc of motion.

Positive finding: Irregularity in movement, pain, patient apprehension.

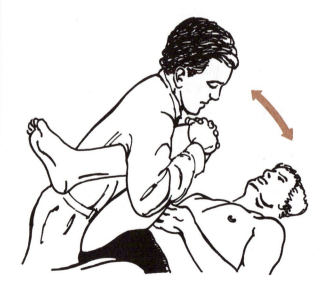

Figure 11.7   Patrick's (Faber) test.

Hip Flexor Length

Assessment: Test the length of hip flexors.

Patient: Supine.

Test position: Patient brings one knee to the chest and maintains a flat lumbar spine. The other leg should remain straight.

Operator: Assess the patient's low back to remain flat and the distance the straight leg is from the table. (If the straight leg lowers while the knee is straight on that side, this indicates tightness in the rectus femoris.)

Positive finding: Straight leg remains a distance from the table (figure 11.8).

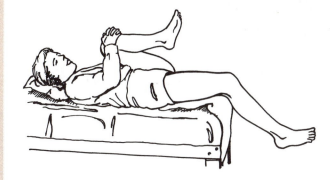

Figure 11.8 Hip flexor length test.

Modified Ober Test

Assessment: To test length of iliotibial band (figure 11.9).

Patient: Side-lying with underlying leg bent.

Test position: Top leg straight with hip extended.

Operator: Stabilize pelvis to prevent lateral tilt, keep leg from internally rotating, and bring it back into extension. Allow leg to drop into adduction.

Positive finding: Leg does not drop into adduction or drops less than 10°.

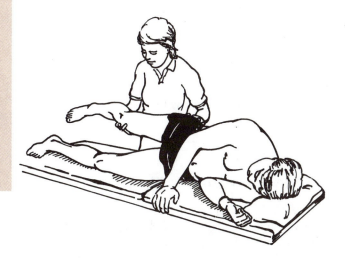

Figure 11.9 Ober's test.

Ely's Test

Assessment: Rectus femoris length (figure 11.10).

Patient: Prone.

Test position: Hip in extension.

Operator: Passively flex knee.

Positive finding: Anterior pelvic tilt indicating tightness.

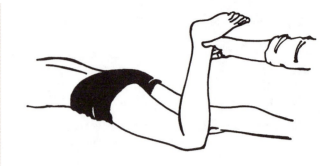

Figure 11.10 Ely's test.

Piriformis

Assessment: To assess piriformis involvement (figure 11.11).

Patient: Prone.

Test position: Knee flexed to 90°.

Operator: Adduct hip, internally rotate hip slightly; apply external rotation force for 20 to 30 s.

Positive finding: Pain.

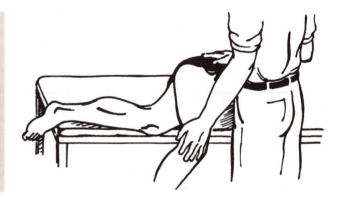

Figure 11.11 Piriformis test.

Trendelenburg's Test

Assessment: Strength of gluteus medius to stabilize pelvis during stance.

Patient: Standing.

Test position: Patient asked to bear weight on one side.

Operator: Assess pelvis level before and during single stance.

Positive finding: Pelvis drop on opposite side of stance leg.

Ortolani's Test

Assessment: Congenital hip dysplasia (figure 11.12).

Patient: Supine.

Test position: Hip in flexion and adduction.

Operator: Press the hips into a posterior direction; relocate hip by abducting the hips and pressing anteriorly.

Positive finding: Click, relocation.

Figure 11.12 Ortolani's sign.

Hamstring Length

Assessment: To assess the length of the hamstring muscle.

Patient: Supine.

Test position: Back neutral, knee extended, hip extended.

Operator: Passively bring the leg up into increasing hip flexion.

Positive finding: Less than 80° of hip flexion indicates tightness in the connective tissue of the posterior thigh.

ARTHROKINEMATICS

Flexion

Nonweight bearing (NWB): The head spins and slightly glides posteriorly and inferiorly.

Weight bearing (WB): The pelvis glides anteriorly on the fixed femur.

Extension

NWB: The head spins and slightly glides anteriorly in extension.

WB: The pelvis glides posteriorly on the fixed femur.

Abduction

NWB: Inferior glide of the convex femoral head. With the hip flexed to 90° the femoral head glides anteriorly in abduction.

WB: In a weight-bearing state, the concave acetabulum glides toward the opposite pelvis.

Adduction

NWB: superior glide. With the hip flexed to 90° the femoral head glides posteriorly in adduction.

Internal Rotation

NWB: Internal rotation is accomplished by posterior glide of the femoral head. With the hip flexed to 90° the femoral head glides inferiorly.

WB: The acetabulum spins about the femoral head toward the side of rotation. For right lower-extremity internal rotation, the pelvis will rotate to the right.

External Rotation

NWB: External rotation is accomplished by anterior glide of the femoral head. With the hip flexed to 90° the femoral head glides superiorly.

WB: The acetabulum spins about the femoral head opposite the side of rotation. For right lower-extremity external rotation, the pelvis will rotate to the left.

NEUROLOGY

Table 11.1 Neurology

Nerve root	Reflex	Motor	Sensory
L2	---	Hip flexion	Anterior proximal thigh
L3	Patellar tendon	Quadriceps	Lateral thigh
L4	Anterior tibialis	Anterior tibialis	Lateral thigh to lateral knee
L5	Proximal hamstring	Extensor hallucis longus	Posterior thigh, dorsum of the foot
S1	Achilles tendon	Peroneus longus	Posterior thigh and lateral foot

Table 11.2 Peripheral Nerves

Nerve	Motor	Sensory
Obturator	Hip adduction and external rotation	Medial inner thigh
Femoral	Hip flexion	Medial thigh (anterior femoral cutaneous)
Tibial	Plantarflex, adduct, or invert the foot	Heel
Superficial peroneal	Eversion of the foot	Lateral calf
Deep peroneal	Dorsiflexion of the foot	Cleft between 1st and 2nd toe

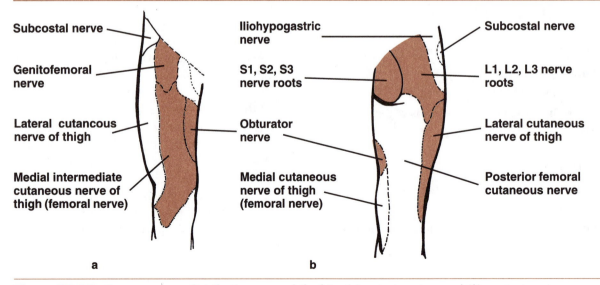

Figure 11.13 Sensory nerve distribution around the hip: (*a*) anterior view and (*b*) posterior view.

SURFACE PALPATION

Greater trochanter
Anterior superior iliac spine
Ischial tuberosity
Inguinal ligament

Femoral triangle
Pubic symphysis
Posterior superior iliac spine

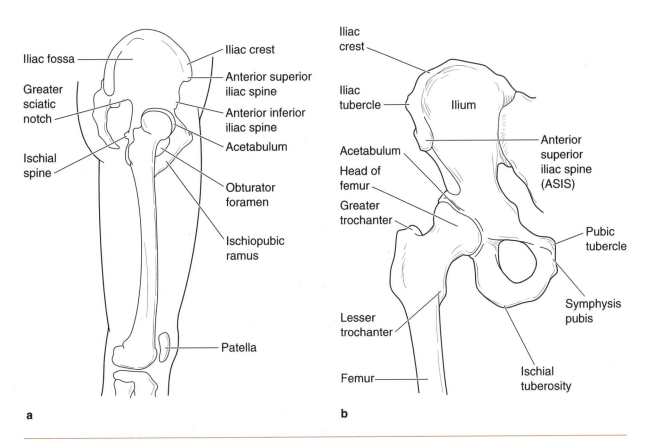

a

b

Figure 11.14   Boney structures of the hip: (*a*) lateral view and (*b*) anterior view.

MUSCLE ORIGIN AND INSERTION

Table 11.3   Muscle Origin and Insertion

Muscle	Origin and insertion
Psoas	Tranverse processes, body L1-L5 and T12 *to* lesser trochanter
Iliacus	Inner surface of ilium and sacrum *to* lesser trochanter
Rectus femoris	Anterior inferior iliac spine *to* patella and tibial tuberosity
Sartorius	Anterior superior iliac spine *to* medial tibia (pes anserine)
Pectineus	Pectineal line on pubis *to* below lesser trochanter *(continued)*

Table 11.3 *(continued)*

Muscle	Origin and insertion
Adductor longus	Inferior rami of pubis *to* middle third of posterior femur
Adductor brevis	Inferior rami of pubis *to* upper half of posterior femur
Gracilis	Inferior rami of pubis *to* medial tibia (pes anserine)
Biceps femoris	Ischial tuberosity *to* lateral condyle of tibia and head of fibula
Semimembranosus	Ischial tuberosity *to* medial condyle of tibia
Semitendinosus	Ischial tuberosity *to* medial tibia (pes anserine)
Gluteus maximus	Posterior ilium, sacrum, coccyx *to* gluteal tuberosity, iliotibial band
Gluteus medius	Anterior lateral ilium *to* lateral surface of greater trochanter
Adductor magnus	Anterior pubis and ischial tuberosity *to* linea aspera on posterior femur, adductor tubercle
Tensor fasciae latae	Anterior superior iliac spine *to* iliotibial tract
Obturator internus	Sciatic notch and margin of obturator foramen *to* greater trochanter
Obturator externus	Pubis, ischium, and margin of obturator foramen *to* upper, posterior femur
Quadratus femoris	Ischial tuberosity *to* greater trochanter
Piriformis	Anterior, lateral sacrum *to* superior greater trochanter
Superior gemellus	Ischial spine *to* greater trochanter
Inferior gemellus	Ischial tuberosity *to* greater trochanter

MUSCLE ACTION AND INNERVATION

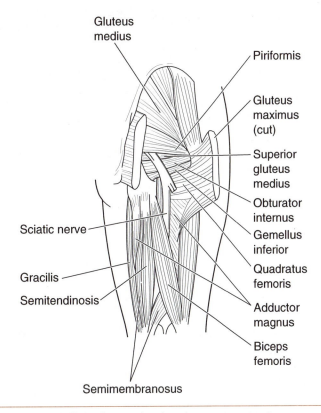

Figure 11.15   Muscles and nerves of the hip and pelvis (posterior view).

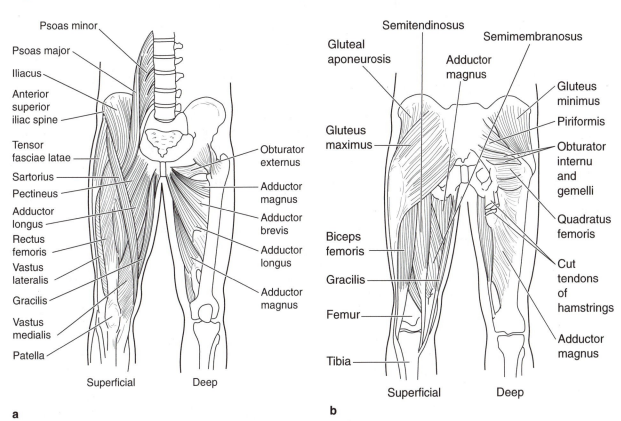

a

b

Figure 11.16   Muscles acting on the hip and pelvis: (*a*) anterior view and (*b*) posterior view.

SECTION
11

Hip

Table 11.4 Muscle Action and Innervation

Action	Muscle involved	Nerve supply	Nerve root
Hip Flexion	Psoas	L1-L3	L1-L3
	Iliacus	Femoral	L2-L3
	Rectus femoris	Femoral	L2-L4
	Sartorius	Femoral	L2-L4
	Pectineus	Femoral	L2-L3
	Adductor longus	Obturator	L2-L4
	Adductor brevis	Obturator	L2-L3, L5
	Gracilis	Obturator	L2-L3
Hip Extension	Biceps femoris	Sciatic	L5, S1-S2
	Semimembranosus	Sciatic	L5, S1-S2
	Semitendinosus	Sciatic	L5, S1-S2
	Gluteus maximus	Inferior gluteal	L5, S1-S2
	Gluteus medius (posterior fibers)	Superior gluteal	L5, S1
	Adductor magnus	Sciatic	L2-L4
Hip abduction	Tensor fasciae latae	Superior gluteal	L4-L5
	Gluteus minimus	Superior gluteal	L5, S1
	Gluteus medius	Superior gluteal	L5, S1
	Gluteus maximus	Inferior gluteal	L5, S1-S2
	Sartorius	Femoral	L2-L3
Hip adduction	Adductor longus	Obturator	L2-L4
	Adductor brevis	Obturator	L2-L4
	Adductor magnus	Obturator	L2-L4
	Gracilis	Obturator	L2-L3
	Pectineus	Femoral	L2-L3
Hip internal rotation	Adductor longus	Obturator	L2-L4
	Adductor brevis	Obturator	L2-L4
	Gluteus minimus (anterior portion)	Superior gluteal	L5, S1
	Gluteus medius (anterior portion)	Superior gluteal	L5, S1
	Tensor fasciae latae	Superior gluteal	L4-L5
	Gracilis	Obturator	L2-L3
	Pectineus	Femoral	L2-L3
Hip external rotation	Gluteus maximus	Inferior gluteal	L5, S1-S2
	Obturator internus	Sacral plexus	L5, S1
	Obturator externus	Obturator	L3-L4
	Quadratus femoris	Sacral plexus	L4-L5, S1
	Piriformis	L5, S1-S2	L5, S1-S2
	Superior gemellus	Sacral plexus	L5, S1
	Inferior gemellus	Sacral plexus	L4-L5, S1
	Sartorius	Femoral	L2-L3
	Gluteus medius	Superior gluteal	L5, S1

DIFFERENTIAL DIAGNOSIS

Table 11.5 Differential Diagnosis

Disorder	Description	Onset	Symptoms	Signs	Special tests
Bursitis	Inflammation of the bursa around the the hip, commonly the ischial, iliopsoas, and trochanteric	Gradual or may be acute from a fall	Local pain and tenderness	Pain with passive and active movements	
Congenital hip dysplasia	Hip dislocation in newborn	Congenital deformity	Hip weakness, balance problems	Limb shortening	Ortolani's test, x-ray
Legg-Calve-Perthes disease	Osteonecrosis of epiphyseal center of proximal femur	3-12 yr old males; 12% bilaterally; family history; insidious	Groin and anterior thigh pain	Limp	X-ray: flattened ossification center
Osteoarthritis	Degenerative changes in femoral head or acetabulum	Gradual wear and tear of the hip joint	Hip pain and stiffness, especially after prolonged positions or activity	Limited ROM in capsular pattern; stiffness	X-ray
Piriformis syndrome	Spasm of the piriformis muscle	Insidious; overuse	Posterior hip pain that may be aggravated by sitting or walking	Local tenderness; limited internal rotation of the hip	Piriformis stretch test
Slipped capital femoral epiphysis	Slippage of the head of the femur at the epiphyseal	11-13 yr old females; 13-16 yr old males; 30% bilaterally; acute or chronic	Hip pain and tenderness, anterior thigh pain	Limited abduction, internal rotation, flexion	X-ray
Avascular necrosis	Deterioration of femoral head	Occurs following severe	Ache, stiffness, pain	Limited ROM; decreased strength	X-ray

(continued)

SECTION
11

Hip

Table 11.5 *(continued)*

Disorder	Description	Onset	Symptoms	Signs	Special tests
Avascular necrosis *(continued)*	secondary to occlusion of of the media/ lateral circumflex artery	trauma to the hip such as fracture or dislocation; may be congenital			
Muscle strain	Microtearing of the muscle or connective tissue about the hip: adductors hamstrings, quadriceps	Acute trauma or overuse	Discomfort locally at site of strain	Local tenderness, decreased muscle power, tissue hiatus	

KNEE

The knee joint is composed of the tibiofemoral joint, the superior tibiofibular joint, and the patellofemoral joint. The knee joint is susceptible to injury because its main stability is via ligamentous support.

TIBIOFEMORAL JOINT

Articulation

A convex medial and lateral condyle of the distal femur and the two concave medial and lateral condyles of the proximal tibia.

Type of Joint

Diarthrodial ginglymus joint.

Degrees of Freedom

- Flexion and extension in the sagittal plane about a horizontal axis through the femoral condyles.
- Internal and external rotation in transverse plane about a vertical axis through the medial intercondylar eminence.
- Abduction and adduction in frontal plane about a sagittal axis throughout the center of the knee.

Active Range of Motion

Flexion: 0-135°
Extension: 0-10° hyperextension
Tibial internal rotation: 0-30°
Tibial external rotation: 0-40°

Accessory Movement

Anterior Glide

Assessment: To assess anterior glide of the tibia on the femur; to facilitate extension.

Patient: Supine with knee flexed to 20° (can also be done prone).

Operator position: Grasp the proximal tibia with one hand and stabilize the distal femur with the other.

Mobilizing force: Glide the tibia in an anterior direction.

Posterior Glide

Assessment: To assess posterior glide of the tibia on the femur (figure 12.1); to facilitate flexion.

Patient: Supine with heel flat and knee bent to 20-30°.

Operator position: Grasp both hands around the proximal tibia.

Mobilizing force: Apply a posterior force to the tibia.

Figure 12.1 Posterior glide of knee joint.

Lateral Glide

Assessment: To assess lateral glide of the tibia on the femur.

Patient: Side-lying with a slight flexed knee and leg extending over edge of the table.

Operator position: One hand placed on proximal tibia with the other hand supporting the leg at the ankle.

Mobilizing force: Apply a lateral force to the tibia with proximal hand.

Medial Glide

Assessment: To assess medial glide of the tibia on the femur.

Patient: Side-lying with a slight flexed knee and leg extending over edge of the table.

Operator position: One hand placed on proximal tibia with the other hand supporting the leg at the ankle.

Mobilizing force: Apply a medial force to the tibia with proximal hand.

Lateral Rotation

Assessment: To assess lateral rotation of the tibia on the femur.

Patient: Sitting with the leg unsupported and the knee at 90°.

Operator position: Grasp both hands around the proximal tibia.

Mobilizing force: Apply a lateral rotation force on the tibia.

Medial Rotation

Assessment: To assess medial rotation of the tibia on the femur.

Patient: Supine with heel flat and knee bent to 20-30°.

Operator position: Grasp both hands around the proximal tibia.

Mobilizing force: Apply a medial rotation force on the tibia.

End-Feel

Flexion: *Soft* due to contact between muscle bulk of posterior calf and thigh; *firm*—due to tension in quadriceps muscle or anterior joint capsule

Extension: *Firm* due to posterior capsule, arcuate complex

Capsular Pattern

Greater limitation of flexion than extension; no rotary restriction.

Close-Packed Position

Maximal extension.

Loose-Packed Position

Midflexion.

Stability

Note the ligaments that stabilize the tibiofemoral joint illustrated in figure 12.2.

Medial

Medial collateral ligament (MCL), anterior cruciate ligament (ACL), medial capsule, posterior cruciate ligament (PCL), meniscofemoral, meniscotibial.

Lateral

Lateral collateral ligament (LCL), lateral meniscus, arcuate ligament, ACL, PCL.

Anterior

ACL, MCL, LCL, quadriceps tendon.

Posterior

PCL, posterior oblique ligament, arcuate complex.

Rotation

MCL, posterior oblique ligament, ACL.

> **MCL:** medial condyle of femur to anteromedial surface of tibial condyle; controls valgus force

SECTION 12

Knee

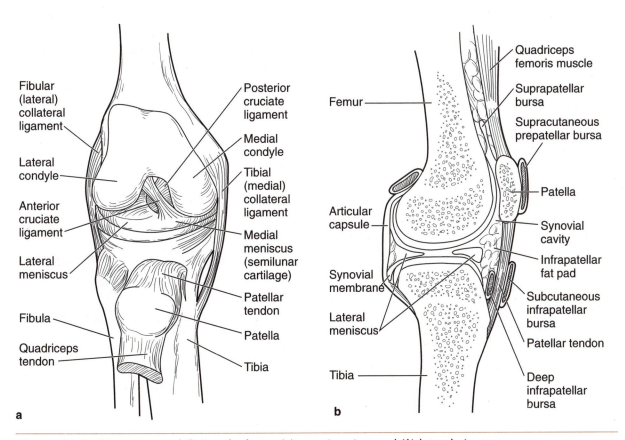

Figure 12.2 Ligaments stabilizing the knee: (*a*) anterior view and (*b*) lateral view.

at the knee; may limit extension, external rotation of tibia on femur.

LCL: lateral condyle of femur to inferior posterior portion of the head of the fibula; controls varus force.

ACL: anterior intercondylar area of the tibia, passing superior, posterior, and laterally to be attached to posterior surface of medial surface of lateral femoral condyle; two bands that are taut throughout range of motion; greatest resistance with extreme hyperextension, limits tibial internal rotation.

PCL: posterior intercondylar area of tibia, passes superiorly/anteriorly and medially to attach to anterior part of medial femoral condyle; two bands—anteromedial, which is taut in flexion, and posterolateral, which is taut in extension; limits tibial internal rotation.

Menisci:

a. Medial: attached to articulating surface of tibial plateau, deep surface of knee joint capsule, and anterior/posterior intercondylar fossa; more C-shaped; thicker posteriorly than anteriorly

b. Lateral: attached to articulating surface of tibial plateau, deep surface of knee joint capsule, and anterior/posterior intercondylar fossa; more O-shaped; equal thickness anterior and posterior; thicker along periphery

Quadriceps and patella tendon: Extension of quadriceps muscle that courses over patella and inserts on tibial tubercle. The patella improves the efficiency of extension because it holds the quadriceps tendon away from the axis of movement.

Arcuate ligament: Fibular head to posterolateral femoral condyle.

Arcuate complex: Arcuate ligament, LCL, popliteus, lateral head of the gastrocnemius.

Oblique popliteal ligament: Bridges the posterior femoral condyles, blends with the semimembranosus, runs with the popliteus muscle,

provides reinforcement to the lateral capsule, and limits anterior medial rotation of the tibia.

Posterior oblique ligament: Posterior to MCL at the posteromedial corner of the capsule.

Special Tests

Anterior Instability

Pivot Shift Test

Assessment: Anterolateral rotary instability of the knee.

Patient: Supine with leg relaxed, hip flexed to 30° and slight internal rotation.

Test position: Hold the patient's foot with one hand while placing the other at the level of the fibular head. The heel of that hand is placed behind the fibula and the knee is extended (this position actually subluxes the tibia anteriorly).

Operator: Apply a valgus force to the knee while maintaining an internal rotation torque on the tibia. The leg is flexed.

Positive finding: At approximately 30-40° the tibia will reduce with a clunk.

Anterior Drawer Sign

Assessment: ACL integrity.

Patient: Supine.

Test position: Knee flexed to 90° and foot flat on the table.

Operator: Sit on the patient's foot and grasp the proximal tibia with both hands. Attempt to pull the tibia forward.

Positive finding: Excessive anterior translation of tibia as compared to the opposite side.

Lachman's Test

Assessment: ACL integrity.

Patient: Supine.

Test position: Knee flexed to 15-20°.

Operator: Grasp the femur with one hand and the tibia with the other hand. Hold the femur steady while exerting an anterior force to the tibia.

Positive finding: Excessive anterior translation of the tibia on the femur as compared to the opposite side (figure 12.3).

Figure 12.3 Lachman's test.

Flexion Rotation Drawer

Assessment: Anterolateral instability of the knee (figure 12.4).

Patient: Supine.

Test position: Hold the patient's ankle between your arm with hands around the tibia. (The weight of the thigh causes the femur to drop posteriorly and externally rotate, producing an anterior subluxation of the lateral tibial plateau.)

Operator: Flex the patient's knee to 20-30° while maintaining a neutral tibia. The tibia is then pushed posteriorly.

Positive finding: Reduction of subluxed tibia.

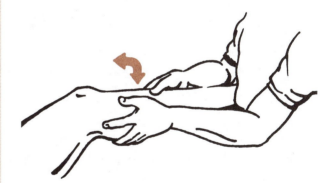

Figure 12.4 Flexion rotation drawer (Noyes) test.

SECTION 12

Knee

Posterior Sag

Assessment: To assess PCL integrity (figure 12.5).

Patient: Supine.

Test position: Hip and knee flexed to 90°.

Operator: Hold the tibia in a horizontal plane.

Positive finding: Tibia sags posteriorly.

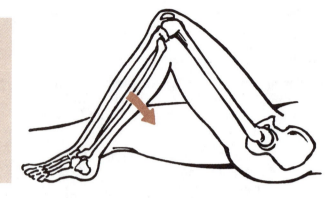

Figure 12.5 Sag sign.

Medial Instability—Valgus Stress Test

Assessment: MCL integrity (figure 12.6).

Patient: Supine.

Test position: Knee flexed to 30°.

Operator: Cradle the test extremity. Place one hand over the lateral knee joint and stabilize the ankle with the other hand. Apply a valgus force to the knee.

Positive finding: Excessive gapping of the medial joint with or without pain. Repeating the test in full extension with a positive finding indicates involvement of the PCL as well as the MCL.

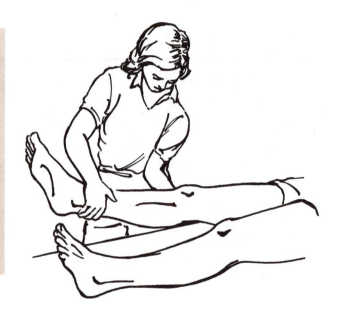

Figure 12.6 Valgus stress test.

Posterior Drawer Sign

Assessment: PCL integrity.

Patient: Supine.

Test position: Knee flexed to 90° and foot flat on the table.

Operator: Sit on the patient's foot and grasp the proximal tibia with both hands. Attempt to push the tibia backward.

Positive finding: Excessive posterior translation of tibia as compared to the opposite side.

Lateral Instability—Varus Stress Test

Assessment: LCL integrity.

Patient: Supine.

Test position: Knee flexed to 30°.

Operator: Cradle the test extremity. Place one hand over the medial knee joint and stabilize the ankle with the other hand. Apply a varus force to the knee.

Positive finding: Excessive gapping of the lateral joint with or without pain. Repeating the test in full extension with a positive finding indicates involvement of the PCL as well as the LCL.

Meniscal Instability

Grind Test

Assessment: Meniscus (figure 12.7).

Patient: Supine.

Test position: Knee flexed in varying degrees of flexion.

Operator: Apply compressive and rotation force through the tibia.

Positive finding: Click and/or pain.

Figure 12.7 Grind test (Bragard's sign).

Apley's Test

Assessment: Meniscal integrity (figure 12.8).

Patient: Prone.

Test position: Knee bent to 90°.

Operator: Apply compressive force through the tibia.

Positive finding: Pain.

Figure 12.8 Apley's test.

PROXIMAL TIBIOFIBULAR JOINT

Articulation

Concave facet on the head of the fibula and the convex facet on the lateral condyle of the tibia.

Type of Joint

Synovial plane gliding joint.

Degrees of Freedom

None; only accessory movement occurs.

Active Range of Motion

None.

Accessory Movement

Anterior/Posterior Glide

Assessment: To assess joint play at the tibiofibular joint. The fibular head must move anteriorly on knee flexion and posteriorly on knee extension.

Patient: Supine with the knee flexed to 90° with foot flat.

Operator position: Stabilize the knee with the medial hand. Grasp the head and neck of the proximal fibula with the lateral hand, the thumb contacting anteriorly and the index and long finger pads contacting posteriorly. (Be cautious of the peroneal nerve.)

Mobilizing force: The lateral hand may glide the proximal fibula posteriorly or anteriorly.

Superior Glide

Assessment: To assess superior glide of the fibula.

Patient: Supine with the knee flexed to 90° with foot flat.

Test position: Have the patient actively dorsiflex the foot.

Operator position: Assess superior motion of the fibular head.

Mobilizing force: This is only an assessment, no force applied.

End-Feel

None.

Capsular Pattern

Pain when joint stressed.

Close-Packed Position

Not applicable.

Loose-Packed Position

Not applicable.

Stability

Anterior Tibiofibular Ligament

From the fibula to anterior tibia, provides anterior stability.

Posterior Tibiofibular Ligament

From the fibula to posterior tibia, provides posterior stability.

Special Tests

None.

PATELLOFEMORAL JOINT

Articulation

Patella articulates with femoral condyles in the trochlear groove.

Type of Joint

Diarthrodial plane joint.

Degrees of Freedom

- Medial and lateral glide in frontal plane.
- Superior and inferior glide in frontal plane.
- Internal and external rotation in frontal plane.

Active Range of Motion

None.

Accessory Movement

Lateral Glide

Assessment: To assess lateral glide.

Patient: Supine with the knee extended and relaxed.

Operator position: Place fingers and thumbs along the medial and lateral borders of the patella.

Mobilizing force: Glide the patella in a lateral direction.

Medial Glide

Assessment: To assess medial glide.

Patient: Supine with the knee extended and relaxed.

Operator position: Place fingers and thumbs along the medial and lateral borders of the patella.

Mobilizing force: Glide the patella in a medial direction.

Superior Glide

Assessment: To assess patellar mobility for knee extension.

Patient: Supine with the knee extended and relaxed.

Operator position: Place the heel of one hand on the inferior aspect of the patella.

Mobilizing force: Glide the patella cranially.

SECTION

12

Knee

Inferior Glide

Assessment: To assess patellar mobility for knee flexion (figure 12.9).

Patient: Supine with the knee extended and relaxed.

Operator position: Place the web space of one hand around the superior border of the patella.

Mobilizing force: Glide the patella caudally.

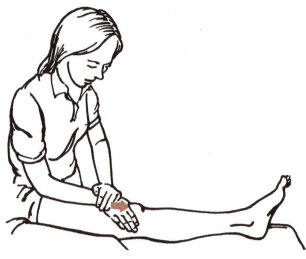

Figure 12.9 Inferior glide of the patellofemoral joint.

End-Feel
Not applicable.

Capsular Pattern
Not applicable.

Close-Packed Position
Full flexion.

Loose-Packed Position
Full extension.

Stability

Retinaculum
Distal connective tissue of quadriceps complex, reinforces patella.

Patellofemoral Ligaments
Inferior pole of patella to tibial tuberosity.

Special Tests

Critical Test

Assessment: To assess patellofemoral pain.

Patient: Sitting.

Test position: Various knee flexion angles from 5° to 90°.

Operator: Position patella in ideal position as patient performs an isometric quadriceps contraction.

Positive finding: Relief of pain indicates faulty patella tracking.

Apprehension Test

Assessment: Lateral patella subluxation or dislocation.

Patient: Supine.

Test position: Knee in full extension.

Operator: Glide patella laterally.

Positive finding: Patient becomes apprehensive and won't allow operator to glide patella.

ARTHROKINEMATICS

Weight Bearing

Knee Extension

Femoral condyles roll anterior and slide posterior; near end extension, the femur rotates internally on tibia, patella slides superiorly and laterally, the menisci follow the femur and move anteriorly.

Knee Flexion

Femoral condyles roll posterior and slide anterior; at beginning of flexion, femur rotates externally on tibia, patella slides inferiorly, the menisci follow the femur and move posteriorly.

Nonweight Bearing

Knee Extension

Tibial condyle slides and rolls anterior. Patella slides superiorly. During last 15-20° of extension, tibia rotates externally on femur. Fibular head moves posteriorly.

Knee Flexion

Tibia glides and rolls posteriorly on femur. Patella slides inferiorly. Internal rotation of tibia on femur during first 15-20° of knee flexion; fibula moves inferiorly and anteriorly.

Tibial Internal Rotation

Posterior horn of lateral meniscus is compressed with tension on the anterior horn; lateral meniscus moves posterior and medial meniscus moves anterior.

Tibial External Rotation

Posterior horn of medial meniscus is compressed with tension on the anterior horn; lateral meniscus moves anterior and medial meniscus moves posterior.

NEUROLOGY

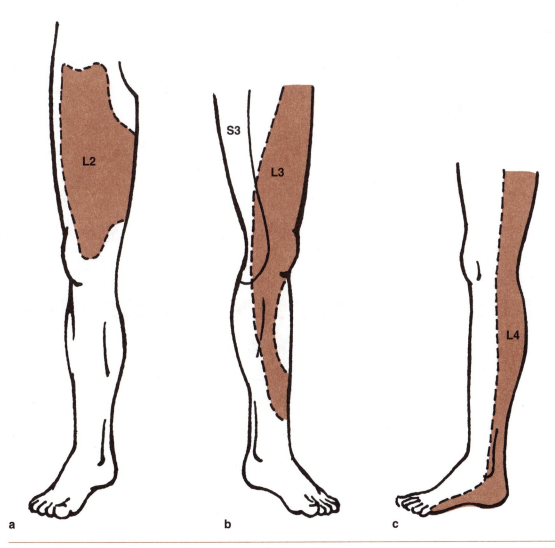

Figure 12.10 Dermatomes about the knee (*a*, *b*, and *c*).

(continued)

SECTION
12

Knee

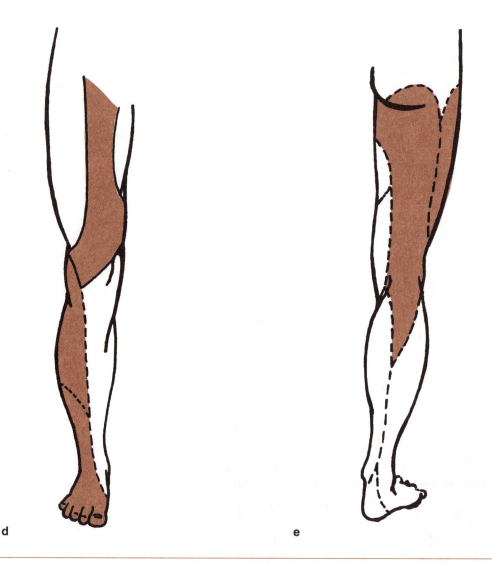

d e

Figure 12.10 Peripheral nerve sensory distribution around the knee (*d*) anterior thigh and calf and (*e*) posterior thigh and calf.

Table 12.1 Neurology

Nerve root	Reflex	Motor	Sensory
L2		Hip flexion	Anterior proximal thigh
L3	Patellar tendon	Quadriceps	Lateral thigh
L4	Anterior tibialis	Anterior tibialis	Medial distal leg
L5	Proximal hamstring	Extensor hallucis longus	Dorsum of the foot
S1	Achilles tendon	Peroneus longus	Lateral foot

Table 12.2 Peripheral Nerves

Nerve	Motor	Sensory
Obturator	Hip adduction and external rotation	Medial inner thigh
Femoral	Hip flexion	Medial thigh (anterior femoral cutaneous)
Tibial	Plantarflex, adduct, or invert the foot	Heel
Superficial peroneal	Eversion of the foot	Lateral calf
Deep peroneal	Dorsiflexion of the foot	Cleft between 1st and 2nd toe

SURFACE PALPATION

Femoral condyles

Tibial tuberosity

Joint line

Pes anserine

Gerdy's tubercle

Tibial plateau

MCL

LCL

Fibular head

Patella

Adductor tubercle

Quadriceps

Popliteal fossa

Gastrocnemius

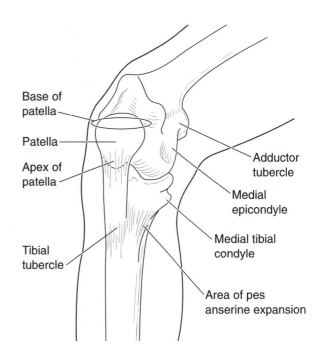

Figure 12.11   Landmarks of the knee.

MUSCLE ORIGIN AND INSERTION

Table 12.3 Muscle Origin and Insertion

Muscle	Origin and insertion
Rectus femoris	Anterior inferior iliac spine *to* patella and tibial tuberosity
Grastrocnemius	Medial and lateral condyles of femur *to* calcaneus
Gracilis	Inferior rami of pubis *to* medial tibia (pes anserine)

(continued)

Table 12.3 *(continued)*

Muscle	Origin and insertion
Popliteus	Lateral condyle of femur *to* proximal tibia
Sartorius	Anterior superior iliac spine *to* medial tibia (pes anserine)
Biceps femoris	Ischial tuberosity *to* lateral condyle of tibia and head of fibula
Semimembranosus	Ischial tuberosity *to* medial condyle of tibia
Semitendinosus	Ischial tuberosity *to* medial tibia (pes anserine)
Vastus Intermedius	Anterior, lateral femur *to* patella and tibial tuberosity
Vastus lateralis	Intertrochanteric line, linea aspera *to* patella and tibial tuberosity
Vastus medialis	Linea aspera, trochanteric line *to* patella and tibial tuberosity

MUSCLE ACTION AND INNERVATION

Table 12.4 Muscle Action and Innervation

Action	Muscle involved	Nerve supply	Nerve root
Knee flexion	Biceps femoris Semimembranosus Semitendinosus Sartorius Gracilis Popliteus Gastrocnemius Tensor fascia latae Plantaris	Sciatic Sciatic Sciatic Femoral Obturator	L5, S1-S2 L5, S1-S2 L5, S1-S2 L2-L4 L2-L3
Knee extension	Rectus femoris Vastus medialis Vastus lateralis Vastus intermedius Tensor fasciae latae	Femoral	L2-L4
Tibial internal rotation	Popliteus Semimembranosus Semitendinosus Sartorius Gracilis	Sciatic Sciatic Femoral Obturator	L5, S1-S2 L5,S1-S2 L2-L4 L2-L3
Tibial external rotation	Biceps femoris	Sciatic	L5, S1-S2

DIFFERENTIAL DIAGNOSIS

Table 12.5 Differential Diagnosis

Disorder	Description	Onset	Symptoms	Signs	Special tests
Bursitis	Inflammation of the bursa about the knee	Repeated friction or acute direct trauma	Pain in area of bursa	Swelling, limited ROM	
ACL	Tear of the ACL, anterior instability of the knee	Acute trauma, may involve pivoting, recurvatum, deceleration	Pain deep in joint	Immediate swelling, instability, tender to palpation	Lachman's anterior drawer, pivot shift
PCL	Tear of the PCL, posterior instability of the knee	Acute trauma, may involve hyperflexion, posterior force to tibia	Pain	Swelling, instability, tender to palpation	Sag test, posterior drawer
MCL	Tear of the MCL, medial instability of the knee	Acute trauma, involves a clipping valgus force	Pain	Swelling, instability, tender to palpation	Valgus stress test
LCL	Tear of the LCL, lateral instability of the knee	Acute trauma, Involves a varus force	Pain	Swelling, instability, tender to to palpation	Varus stress test
Meniscal injury	Tear of the semilunar cartilage	Acute trauma, involving compression and rotation of the knee	Pain, inability to squat	Swelling, buckling, tender at joint line	Grind test Apley's test
Osgood-Schlatter's disease	Dysfunction of apophyseal area of tibial tubercle	males 10-14 yr old, increased activity, may be associated with sudden growth in height	Pain with quadriceps contraction, squatting along the tibial tubercle	Swelling and deformity at the tibial tubercle	
Patello femoral pain	Dysfunction of patello-femoral joint, may involve articular cartilage destruction	Usually chronic in nature, more common in in females	Pain about the patella or on the under-surface, stiffness	Tender to palpation, uncomfortable with squats, stair climbing	Critical test

(continued)

SECTION **12** Knee

Table 12.5 *(continued)*

Disorder	Description	Onset	Symptoms	Signs	Special tests
Patellar dislocation	Disruption of the patello-femoral joint where the patella disarticulates with the femur, usually laterally	Acute trauma with forceful quadriceps contraction and valgus force, congenital	Pain	Obvious deformity when dislocated, swelling	Apprehension test
Popliteal cyst (Baker's cyst)	Herniation of the synovium	May be associated with meniscal injury	Pain posterior knee	Swelling popliteal area	
Osteo-chondritis dissecans	Loose bodies in the joint, mainly found on the posterolateral aspect	Trauma, impairment of blood supply, heredity	Knee pain	Swelling, quadriceps atrophy, tenderness to palpation	X-ray

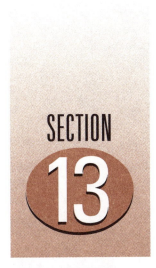

FOOT AND ANKLE

The foot and ankle involve the distal tibiofibular joint, talocrural joint, subtalar joint, and transtarsal, tarsometatarsal, metatarsalphalangeal, and interphalangeal joints in the toes as well as the midfoot in which no active motion occurs.

DISTAL TIBIOFIBULAR JOINT (DTF)

Articulation

The convex lower end of the fibula and the concave fibular notch at the distal end of the tibia.

Type of Joint

Synarthrosis syndesmosis.

Degrees of Freedom

None.

Active Range of Motion

No active motion.

Accessory Movement

Posterior Glide

Assessment: To assess accessory motion of dorsiflexion.

Patient: Supine with knee in extension.

Operator position: Place fingers of medial hand under the tibia and the thumb over the tibia to stabilize it. Place the lateral hand using the thenar eminence over the lateral malleolus, with fingers underneath.

Mobilizing force: Glide the lateral malleolus posteriorly, directing force through the left thenar eminence.

End-Feel

Not applicable.

Capsular Pattern

Pain when joint stressed.

Close-Packed Position

Not applicable.

Loose-Packed Position

Not applicable.

Stability

Anterior and Posterior Tibiofibular Ligaments

Connect tibia and fibula together anterior and posterior to interosseous ligament.

Crural Interosseous Tibiofibular Ligament

Connects the tibia and fibula and is continuous with the interosseous membrane.

Special Tests

None.

TALOCRURAL JOINT (TC)

Accessory Movement

Distraction

Assessment: To increase joint play at the ankle mortise. Oscillations may also be used for pain control (figure 13.1).

Patient: Supine with lower extremity extended and ankle in a resting position.

Operator position: Stand or sit at end of table and wrap the fingers of both hands over the dorsum of the patient's foot. Place your thumbs on the plantar surface of the foot.

Mobilizing force: Distract the joint with both hands by leaning back, being careful to keep patient's ankle in mobilizing plane.

Articulation

The tibia, fibula, and talus; the superior convex dome of the talus fits into the concave surface formed by the medial malleolus, distal tibial, and lateral malleolus.

Type of Joint

Diarthrodial hinge joint.

Degrees of Freedom

Dorsiflexion and plantarflexion in sagittal plane about a coronal axis that passes approximately through fibular malleolus and the body of the talus; forms an 80° angle from vertical; lateral malleolus slightly posterior and inferior to medial malleolus.

Active Range of Motion

Dorsiflexion: 0-20°
Plantarflexion: 0-50°

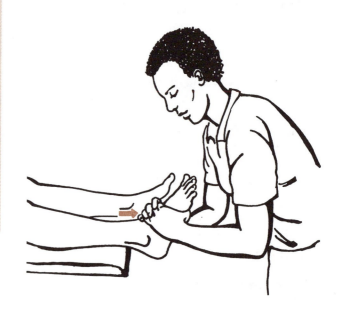

Figure 13.1 Distraction of the talocrural joint.

Ventral Glide

Assessment: To assess plantarflexion (figure 13.2).

Patient: Prone with knee relaxed and foot over the edge of the table.

Operator position: Your cranial hand grasps the anterior and distal surface of the tibia and fibula. The caudal hand contacts the posterior talus and calcaneus with the web space.

Mobilizing force: Glide the calcaneus and talus downward in an anterior direction.

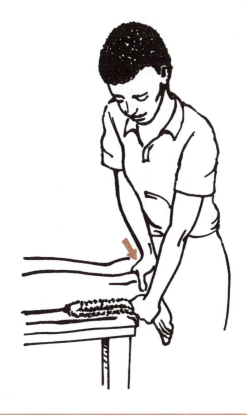

Figure 13.2 Ventral glide of the talocrural joint.

Dorsal Glide

Assessment: To assess dorsiflexion (figure 13.3).

Patient: Supine and relaxed with heel over the edge of the table.

Operator position: Stabilize the distal tibia against the plinth with the cranial hand, wrapping the fingers around posteriorly. The web of the caudal hand grasps the neck of the talus with the fingers wrapped around the foot.

Mobilizing force: Glide the talus posteriorly on the tibia.

Figure 13.3 Dorsal glide of the talocrural joint.

End-Feel

Dorsiflexion: *Firm* due to tension in posterior capsule, Achilles tendon, posterior portion of deltoid and calcaneofibular ligament, and posterior talofibular ligament

Plantarflexion: *Firm* due to tension in anterior capsule, anterior portion of deltoid and anterior talofibular ligament, anterior tibial muscle, and long extensors of the toes

Plantarflexion: *Hard* due to posterior tubercle of talus contacting posterior tibia

Capsular Pattern

Plantarflexion limited more than dorsiflexion.

Close-Packed Position

Maximum dorsiflexion.

Loose-Packed Position

10° of plantarflexion midway between inversion and eversion.

Stability

Figure 13.4 illustrates the ligaments that stabilize the talocrural joint.

Medial Ligaments

Deltoid: consists of the posterior tibiotalar, tibiocalcaneal, tibionavicular, and anterior

Posterior tibiotalar: from the medial malleolus to the posterior talus process.

Tibiocalcaneal: from medial malleolus to sustentaculum tali.

Tibionavicular: medial malleolus to navicular bone and spring ligament.

Anterior tibiotalar: medial malleolus to the navicular bone; stabilizes medial ankle.

Lateral Ligaments

Anterior talofibular ligament: lateral malleolus to the lateral talus.

Calcaneofibular ligament: tip of lateral malleolus downward and backward to the lateral surface of calcaneus.

Posterior talofibular ligament: lateral malleolus to the posterior tubercle of the talus; stabilizes lateral ankle.

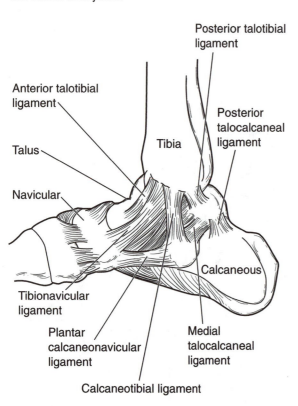

a Medial aspect

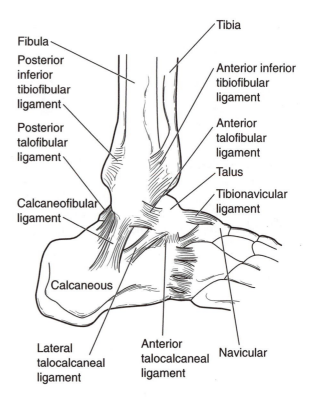

b Lateral aspect

Figure 13.4 Ligaments supporting the foot and ankle.

Special Tests

Lateral Instability

Assessment: To test the integrity of the anterior talofibular ligament.

Patient: Supine.

Test position: Foot in slight plantarflexion.

Operator: Grasping the foot, apply an inversion force to the TC and subtalar joint.

Positive finding: Lateral gapping and/or pain as compared to the opposite leg.

Anterior Instability

Assessment: To test the integrity of the anterior tibiofibular ligament.

Patient: Sitting.

Test position: Foot in slight plantarflexion.

Operator: Grasp the posterior calcaneus with one hand and the anterior distal tibia with the other. Apply an anterior force to the calcaneus.

Positive finding: Excessive anterior movement of the talus as compared to the opposite leg.

Medial Instability

Assessment: To test the integrity of the deltoid ligament.

Patient: Supine.

Test position: Foot in slight plantarflexion.

Operator: Grasping the foot, apply an eversion force to the TC and subtalar joint.

Positive finding: Medial gapping and/or pain as compared to the opposite leg.

Achilles Tendon: Thompson's Test

Assessment: To test the integrity of the Achilles tendon.

Patient: Prone.

Test position: Feet over the edge of a table.

Operator: Squeeze the calf at the middle of the muscle belly.

Positive finding: A normal response would be plantarflexion of the foot. If the tendon is ruptured, this movement is markedly decreased or absent.

SUBTALAR JOINT (STJ; TALOCALCANEAL JOINT)

Articulation

The superior talus and inferior calcaneus; posteriorly, the concave facet on the inferior surface of the talus and convex facet on the body of the calcaneus. The anterior and middle articulations are formed by two convex facets on the talus with two concave facets on the calcaneus.

Type of Joint

Diarthrodial bicondylar joint with triplane motion: motion consisting of inversion and eversion (frontal plane), abduction and adduction (transverse plane), plantarflexion and dorsiflexion (sagittal plane).

Degrees of Freedom

- Pronation and supination in three planes.
- Axis of motion—oblique axis from posterolateral plantar aspect to anteromedial dorsal aspect.

Active Range of Motion

Inversion: 0-30°
Eversion: 0-10°

Accessory Movement

SECTION

13

Foot and Ankle

Distraction

Assessment: General mobility, pain control (figure 13.5).

Patient: Supine with leg supported on the table.

Operator position: The distal hand grasps around the calcaneus from the posterior aspect of the foot. The other hand fixates the talus.

Mobilizing force: Pull the calcaneus distally with respect to the long axis of the leg.

Figure 13.5 Distraction of the subtalar (talocalcaneal) joint.

Medial Glide

Assessment: To assess inversion (figure 13.6).

Patient: Side-lying with leg supported on the table and lateral side up.

Operator position: Align shoulder and arm parallel to the bottom of the foot. Stabilize the talus with your proximal hand. Place the base of the distal hand on the side of the calcaneus laterally and wrap the fingers around the plantar surface.

Mobilizing force: Apply a glide medially.

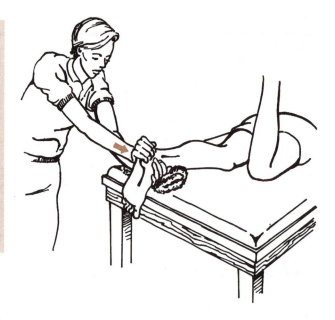

Figure 13.6 Medial glide of the subtalar joint.

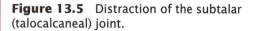

Lateral Glide

Assessment: To assess eversion (figures 13.7).

Patient: Side-lying with leg supported on the table and the medial side up.

Operator position: Align shoulder and arm parallel to the bottom of the foot. Stabilize the talus with your proximal hand. Place the base of the distal hand on the side of the calcaneus medially and wrap the fingers around the plantar surface.

Mobilizing force: Apply a glide laterally.

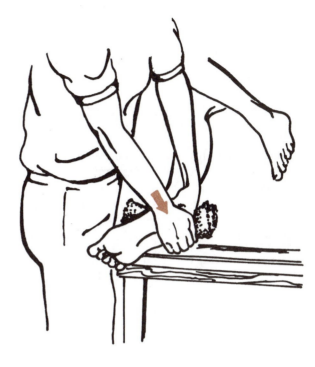

Figure 13.7 Lateral glide of the subtalar joint.

End-Feel

Inversion: *Firm* (lateral joint capsule and lateral ligaments)

Eversion: *Firm* (joint capsule, deltoid ligament, and posterior tibialis muscle)

Eversion: *Hard* (calcaneus and sinus tarsi)

Capsular Pattern

Supination limited more than pronation; inversion limited more than eversion.

Close-Packed Position

Supination.

Loose-Packed Position

Pronation.

Stability

Cervical Talocalcaneal Ligament

From inferolateral aspect of talar neck downward and lateral to dorsum of calcaneus; restricts inversion.

Interosseus Talocalcaneal Ligament

From underside of talus at sustentaculi tali downward and lateral to dorsum of calcaneus; restricts eversion.

Special Tests

STJ Neutral

Assessment: Assessing STJ neutral (figure 13.8).

Patient: Lies prone with feet over the end of the table.

Test position/Operator: Grasp the patient's foot over the fourth and fifth metatarsal heads with index finger and thumb; the other hand is palpating both sides of the talus on the dorsum of the foot using the thumb and index finger. Passively dorsiflex to resistance, supinate and pronate foot until talar head does not bulge laterally or medially; assess rear foot and forefoot position.

Positive finding: Rear foot position greater than 2° and forefoot position greater than 2°.

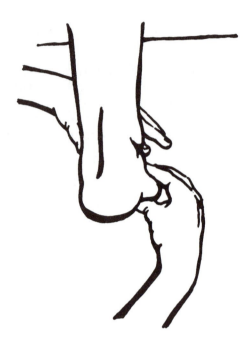

Figure 13.8 Assessing eversion of the subtalar joint.

Navicular Drop Test

Assessment: Excessive STJ pronation.

Patient: Standing.

Test position: Mark the navicular tubercle and then measure in millimeters the vertical distance between the navicular tubercle and the floor.

Operator: Measure twice, once with the STJ in neutral and once with the foot in the natural standing posture of the patient.

Positive finding: Greater than 10 mm difference between two readings indicates excessive STJ pronation.

TRANSTARSAL JOINT (JOINT OF CHOPART, MIDTARSAL JOINT)

Articulation

The calcaneus and cuboid and the talus and navicular. The talonavicular joint is composed of the large convex head of the talus and the concave posterior portion of the navicular bone. The calcaneocuboid joint is composed of the shallow convex (proximal, distal)-concave (medial, lateral) surfaces on the anterior calcaneus and the convex (medial, lateral)-concave (proximal, distal) surfaces on the posterior cuboid (figure 13.9)

Type of Joint

Calcaneocuboid is a sellar-shaped joint; talonavicular joint is a condyloid joint.

Degrees of Freedom

Pronation and supination consisting of:

- Inversion and eversion in the frontal plane,
- Abduction and adduction in the transverse plane, and
- Plantarflexion and dorsiflexion in the sagittal plane.

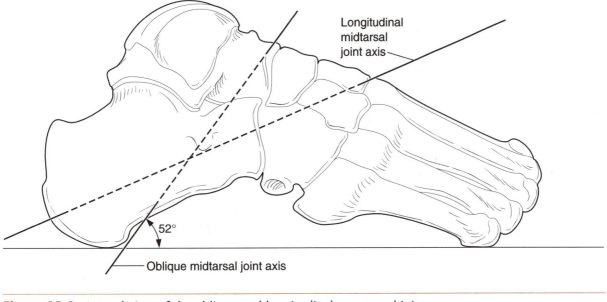

Longitudinal
midtarsal
joint axis

52°

Oblique midtarsal joint axis

Figure 13.9 Lateral view of the oblique and longitudinal transtarsal joint axes.

Active Range of Motion

Longitudinal (8° of motion): 15° upward from transverse plane, 16° medially from longitudinal reference

Oblique (22° of motion): 52° upward from transverse plane and 64° medially from longitudinal foot reference

Accessory Movement

Plantar Glide

Assessment: To increase the arch.

Patient: Supine with the knee relaxed.

Operator position: Fixate the more proximal bone with your finger by grasping dorsally at the level of the talar neck; the thumb wraps around laterally and the rest of the fingers medially. The other hand grasps the mobilizing bone, the thumb contacts dorsally, and the hand and fingers wrap around the foot medially and plantarly.

Mobilizing force: Move the distal bones plantarly.

Dorsal Glide

Assessment: To decrease the arch.

Patient: Prone with hip and knee flexed.

Operator position: Fixate the calcaneus with one hand. With the other hand, wrap your fingers around the lateral side of the foot.

Mobilizing force: Move the distal bones in a dorsal direction.

End-Feel

Supination: *Firm* (lateral joint capsule and lateral ligaments)

Pronation: *Firm* (joint capsule, deltoid ligament, and posterior tibialis muscle)

Pronation: *Hard* (calcaneus and sinus tarsi)

Capsular Pattern

Dorsiflexion more limited than plantarflexion.

Close-Packed Position

Dorsiflexion.

Loose-Packed Position

Plantarflexion.

Stability

Plantar Calcaneonavicular (Spring Ligament)

Attached from the anterior margin of the sustentaculum tali to the inferior surface and tuberosity of the navicular bone; reinforces the medial arch.

Bifurcated (Y Ligament of the Foot)

Stem attached to the upper surface of the anterior portion of the calcaneus, with lateral limb attached to the upper surface of the cuboid and medial limb attached to the upper navicular bone; reinforces laterally.

Long Plantar

Attached from the undersurface of the calcaneus behind and to the undersurface of the cuboid and the bases of the third, fourth, and fifth metatarsal bones in front.

Short Plantar

Attached from the anterior tubercle on the undersurface of the calcaneus and to the adjoining part of the cuboid bone.

Special Tests

None.

MIDFOOT

Articulation

Naviculocuboid, naviculocuneiform, intercuneiform joint.

Type of Joint

Plane synovial joints; cuneonavicular—plane synovial joints; cuboideonavicular and cunicuboid—fibrous joint.

Degrees of Freedom

No active motion occurs.

Active Range of Motion

None.

Accessory Movement

Plantar Glide

Assessment: To increase the arch.

Patient: Supine with the knee relaxed.

Operator position: Fixate the more proximal bone with your finger by grasping dorsally at the level of the talar neck; the thumb wraps around laterally and the rest of the fingers medially. The other hand grasps the mobilizing bone, the thumb contacts dorsally, and the hand and fingers wrap around the foot medially and plantarly.

Mobilizing force: Move the distal bones plantarly.

Dorsal Glide

Assessment: To decrease the arch.

Patient: Prone with hip and knee flexed.

Operator position: Fixate the calcaneus with one hand. With the other hand, wrap your fingers around the lateral side of the foot.

Mobilizing force: Move the distal bones in a dorsal direction.

End-Feel

Not applicable.

Capsular Pattern

Not applicable.

Close-Packed Position

Supination.

Loose-Packed Position

Pronation.

Stability

Plantar Calcaneonavicular (Spring Ligament)

Attached from the anterior margin of the sustentaculum tali to the inferior surface and tuberosity of the navicular bone; reinforces the medial arch.

Bifurcated (Y Ligament of the Foot)

Stem attached to the upper surface of the anterior portion of the calcaneus, with lateral limb attached to the upper surface of the cuboid and medial limb attached to the upper navicular bone; reinforces laterally.

Long Plantar

Attached from the undersurface of the calcaneus behind and to the undersurface of the cuboid and the bases of the third, fourth, and fifth metatarsal bones in front.

Short Plantar

Attached from the anterior tubercle on the undersurface of the calcaneus and to the adjoining part of the cuboid bone.

Special Tests

None.

FOREFOOT (TARSOMETATARSAL JOINTS)

Articulation

Five tarsometatarsal joints. Concave base of the first metatarsal articulates with the convex surface of the medial cuneiform. The bases of the second and third metatarsals articulate with the mortise formed by the intermedial cuneiform and the sides of the medial and lateral cuneiforms. The base of the third metatarsal articulates with the lateral cuneiform, and the bases of the fourth and fifth metatarsals articulate with the cuboid.

Type of Joint

Tarsometatarsal and intermetatarsal joint (first ray): diarthrodial plane joint.

Degrees of Freedom

- Flexion and extension in sagittal plane about a coronal axis.
- Abduction and adduction in the frontal plane about a sagittal axis.

Active Range of Motion

Each ray has its own oblique axis.

Accessory Movement

Distraction

Assessment: To assess mobility.

Patient: Sitting with leg supported.

Operator position: Stabilize the tarsal with one hand; grasp with the thumb dorsally and index finger plantarly. The other hand grasps around the proximal portion of the metatarsal.

Mobilizing force: Apply long-axis distraction to the metatarsal to separate the joint.

Plantar/Dorsal Glide

Assessment: To assess plantar/dorsal glide of tarsometatarsal joint (TMT joint) for flexion and extension (figure 13.10).

Patient: Lies prone with feet over the end of the table.

Operator position: Stabilize the tarsal with one hand; grasp with the thumb dorsally and index finger volarly. The other hand grasps around the proximal portion of the metatarsal.

Mobilizing force: The thumb on the dorsum of the metatarsal glides the proximal portion of the bone in a volar/dorsal direction.

Figure 13.10 Dorsal gliding of a tarsometatarsal joint.

End-Feel

Firm end-feel in all planes due to ligamentous constraints.

Capsular Pattern

Equal limitation in all directions.

Close-Packed Position

Not defined.

Loose-Packed Position

Not defined.

Stability

Dorsal Ligaments

From the tarsals to the metatarsals; support the arch.

Plantar Ligaments

From anterior margin of calcaneus to the inferior surface of the navicular; support the arch.

Interosseous Ligaments

From the undersurface of the talus to the superior surface of the calcaneus; reinforce foot and limit pronation, supination, and abduction.

Special Tests

None.

METATARSALPHALANGEAL JOINTS (MTP)

Articulation

Formed proximally by the convex heads of the five metatarsals and distally by the concave bases of the proximal phalanges.

Type of Joint

Diarthrodial condyloid joint.

Degrees of Freedom

- Flexion and extension about an oblique axis.
- Abduction and adduction not defined.

Active Range of Motion

Flexion: 0-20°

Extension: 0-70°

Abduction: 0-10°

Accessory Movement

Distraction

Assessment: To assess mobility.

Patient: Sitting with the leg supported.

Operator position: Stabilize the metatarsal with one hand; grasp with the thumb dorsally and index finger plantarly. The other hand grasps around the proximal portion of the proximal phalanx.

Mobilizing force: Apply long-axis distraction to the proximal phalanx to separate the joint.

Plantar/Dorsal Glide

Assessment: To assess plantar/dorsal glide of MTP for flexion and extension.

Patient: Sitting with leg supported.

Operator position: Stabilize the metatarsal with one hand; grasp with the thumb dorsally and index finger volarly. The other hand grasps around the proximal portion of the phalanx.

Mobilizing force: The thumb on the dorsum of the metatarsal glides the proximal portion of the bone in a volar/dorsal direction.

End-Feel

Flexion: *Firm* due to tension in dorsal joint capsule, collateral ligaments, and short toe extensors

Extension: *Firm* due to tension in plantar joint capsule, short toe flexors, and plantar fascia

Abduction: *Firm* due to tension in joint capsule, collateral ligaments, plantar interosseous, and adductor muscle fascia

Capsular Pattern

Greater limitation in extension than flexion 2-5: variable.

Close-Packed Position

Full extension.

Loose-Packed Position

10° of extension.

Stability

Plantar Aponeurosis

From inferior calcaneus, blends with metatarsal heads, reinforces arch of the foot; responsible for windlass effect.

Fibrous Capsule

Surrounds joint and is strengthened by plantar and collateral ligaments.

Deep Transverse Metatarsal Ligament

Blends with plantar ligament, connects the MTP.

Special Tests

None.

INTERPHALANGEAL JOINTS (IP)

Articulation

Concave base of the distal phalanx and convex head of proximal phalanx.

Type of Joint

Diarthrodial hinge joint.

Degrees of Freedom

Flexion and extension in the sagittal plane about a coronal axis.

Active Range of Motion

PIP flexion: 0-90°

DIP flexion: 0-40°

Accessory Movement

Distraction

Assessment: To assess mobility of the joint.

Patient: Leg and foot resting on the treatment table.

Operator position: Fixate the proximal bone with the fingers; wrap the fingers and thumb of your other hand around the distal bone close to the joint.

Mobilizing force: Apply long-axis traction to separate the joint surfaces.

Glides

Volar glide
Assessment: To assess flexion.

Dorsal glide
Assessment: To assess extension.

Patient: Sitting with leg supported.

Operator position: Fixate the proximal bone with the fingers; wrap the fingers and thumb of your other hand around the distal bone close to the joint.

Mobilizing force: The glide force is applied by the thumb against the proximal end of the bone.

End-Feel

PIP flexion: *Soft* due to soft tissues of plantar surfaces contacting each other

PIP flexion: *Firm* due to tension in dorsal joint capsule and collateral ligaments

PIP extension: *Firm* due to tension in plantar joint capsule and plantar fascia

DIP flexion: *Firm* due to tension in dorsal joint capsule, collateral ligaments, and oblique retinacular ligament

DIP extension: *Firm* due to tension in plantar joint capsule and plantar fascia

Capsular Pattern

Flexion limited more than extension.

Close-Packed Position

Full flexion.

Loose-Packed Position

Slight flexion.

Stability

Same as with MTP.

Special Tests

None.

Table 13.1 Range of Motion

Motion	Range of motion (AAOS)
Plantarflexion	0-50°
Dorsiflexion	0-20°
Inversion	0-35°
Eversion	0-15°
Subtalar joint inversion	0-30°
Subtalar joint eversion	0-10°
Great toe extension	0-70°

ARTHROKINEMATICS

Nonweight Bearing

Dorsiflexion

DTF: Fibula rotates laterally and glides proximally to accommodate wider portion of talus engaging the mortise.

TC: Talus slides posteriorly and abducts on tibia wedging into mortise.

STJ: Pronation.

Plantarflexion

DTF: Fibula rotates medially, narrowing mortise as talus disengages from the mortise.

TC: Talus slides anteriorly on tibia and disengages from the mortise.

STJ: Supination.

Inversion

STJ: Calcaneus slides laterally on fixed talus.

TMT: Navicular slides medially and dorsally on talus.

Eversion

STJ: Calcaneus slides medially on fixed talus.

TMT: Navicular slides laterally and toward plantar surface on the talus.

Pronation

Occurs at the STJ and involves calcaneal dorsiflexion, abduction, and eversion.

Supination

Occurs at the STJ and involves calcaneal plantarflexion (coronal axis), adduction (vertical axis), and inversion (longitudinal axis through the foot).

First MTP flexion: plantar glide of base of phalange on the heads of the metatarsal

First MTP extension: dorsal glide

2-5 MTP flexion: plantar glide

2-5 MTP extension: dorsal glide

Abduction

Concave base of phalanges slide on convex heads of metatarsals in lateral direction away from the second toe.

IP Flexion

Concave base of distal phalanx slides on the convex head of the proximal phalanx in the same direction as the shaft of the distal bone; concave base slides toward plantar surface of the foot IP extension: concave base slides toward dorsum of foot during extension.

Weight Bearing

Dorsiflexion

DTF: Fibula glides anterior to greater extent than tibia on talus, results in internal rotation of tibia.

TC: Talus glides posteriorly and medially.

STJ: Pronation.

Plantarflexion

DTF: Fibula glides posterior to greater extent than tibia on talus, results in external rotation of tibia.

TC: Talus glides anteriorly and laterally.

STJ: Supination.

Pronation

Occurs at the STJ and involves adduction and plantarflexion of talus and eversion of calcaneus; when talocalcaneal joint is pronated, two sets of axes at the midtarsal joint are parallel and allow maximal amount of motion.

Supination

Occurs at the STJ and involves abduction and dorsiflexion of talus and inversion of calcaneus; when talocalcaneal is supinated, two sets of axes at the midtarsal joint are divergent and little motion is allowed.

NEUROLOGY

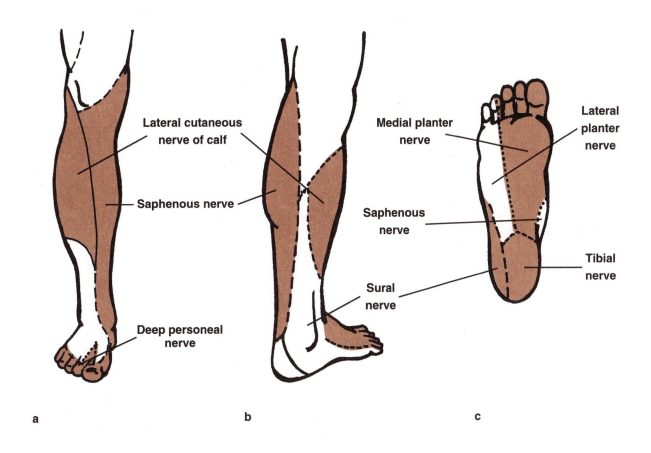

Figure 13.11 Peripheral nerve distribution in the lower leg, ankle, and foot—(*a*) anterior and (*b*) posterior views—and in the foot (*c*).

Table 13.2 Neurology

Nerve root	Reflex	Motor	Sensory
L3	Patellar tendon	Quadriceps	Lateral thigh
L4	Anterior tibialis	Anterior tibialis	Medial distal leg
L5	Proximal hamstring	Extensor hallucis longus	Dorsum of the foot
S1	Achilles tendon	Peroneus longus	Lateral foot

Table 13.3   Peripheral Nerves

Nerve	Motor	Sensory
Obturator	Hip adduction and external rotation	Medial inner thigh
Femoral	Hip flexion	Medial thigh (anterior femoral cutaneous)
Tibial	Plantarflex, adduct, or invert the foot	Heel
Superficial peroneal	Eversion of the foot	Lateral calf
Deep peroneal	Dorsiflexion of the foot	Cleft between 1st and 2nd toe

SURFACE PALPATION

Navicular tubercle

Extensor digitorum longus

Cuboid

Extensor hallucis longus

Peroneal tubercle

Extensor digitorum brevis

Peroneus longus

Dorsal pedal pulse

Peroneus brevis

Anterior tibialis tendon

Peroneus tertius tendon

MUSCLE ORIGIN AND INSERTION

Table 13.4   Muscle Origin and Insertion

Muscle	Origin and insertion
Abductor digiti minimi	Lateral calcaneus *to* base of proximal phalanx of 5th toe
Abductor hallucis	Medial calcaneus *to* medial base of proxima phalanx of 1st toe
Adductor hallucis	2nd, 3rd, and 4th metatarsal *to* lateral side of proximal palanx of big toe
Dorsal interossei	Sides of metatarsals *to* lateral side of proximal phalanx
Extensor digitorum brevis	Lateral calcaneus *to* proximal phalanx of 1st, 2nd, and 3rd toes
Extensor digitorum longus	Lateral condyle of tibia, fibula, interossei membrane *to* dorsal expansion of toes 2-5
Extensor hallucis longus	Anterior fibula, interosseous membrane *to* distal phalanx of big toe

Table 13.4 *(continued)*

Muscle	Origin and insertion
Flexor digiti minimi brevis	5th metatarsal *to* proximal phalanx of little toe
Flexor digitorum brevis	Medial calcaneus *to* middle phalanx of toes 2-5
Flexor digitorum longus	Posterior tibia *to* distal phalanx of toes 2-5
Flexor hallucis brevis	Cuboid *to* medial side of proximal phalanx of big toe
Flexor hallucis longus	Lower 2/3 of posterior fibula and interosseous membrane *to* the plantar aspect of the base of the distal phalanx of the great toe
Gastrocnemius	Medial and lateral condyles of femur *to* posterior calcaneus
Lumbricales	Tendon of FDL *to* base of proximal phalanx of toes 2-5
Peroneus brevis	Lower lateral fibula *to* 5th metatarsal
Peroneus longus	Lateral condyle of tibia and upper, lateral fibula *to* 1st cuneiform, lateral 1st metatarsal
Peroneus tertius	Lower, anterior fibula, interosseous membrane *to* base of 5th metatarsal
Plantar interossei	Medial side of 3rd-5th metatarsal *to* medial side of proximal phalanx of toes 3-5
Plantaris	Linea aspera of femur *to* calcaneus
Quadratus plantae	Medial, lateral inferior calcaneus *to* flexor digitorum tendon
Soleus	Upper posterior tibial fibula interosseous membrane *to* posterior calcaneus
Tibialis anterior	Upper lateral tibia, interosseous membrane *to* medial, plantar surface of 1st cuneiform
Tibialis posterior	Upper posterior tibia, fibula, interosseus membrane *to* inferior navicular

SECTION
13

Foot and Ankle

MUSCLE ACTION AND INNERVATION

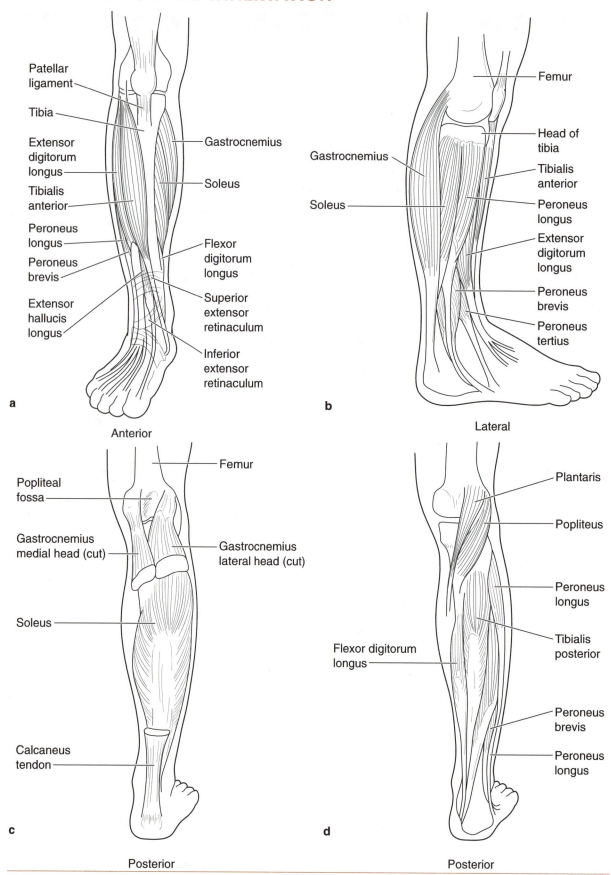

Figure 13.12   Lower leg muscles: (*a*) anterior view, (*b*) lateral view, (*c*) posterior view of superficial aspect, and (*d*) posterior view of deep aspect.

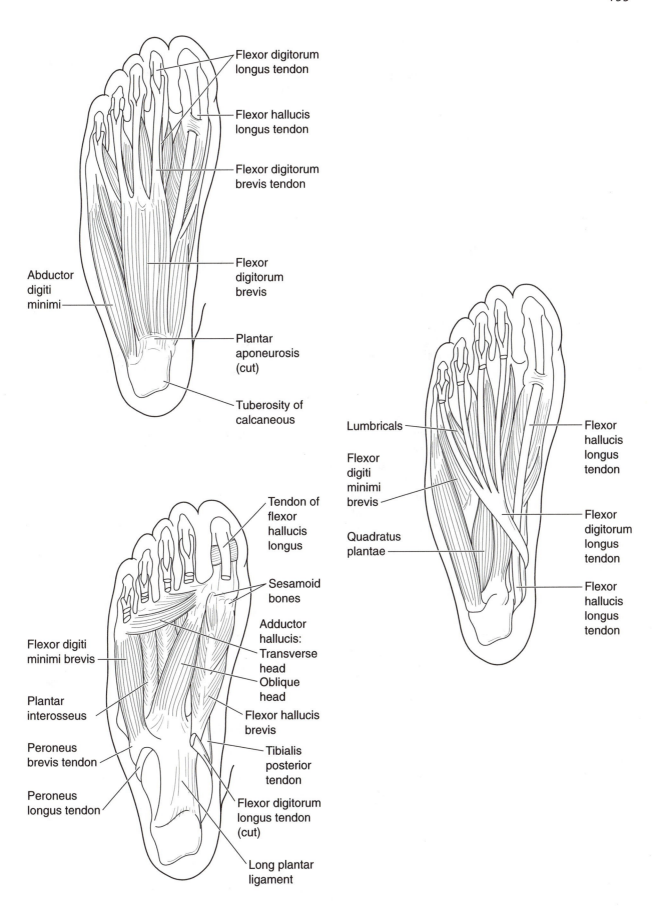

Flexor digitorum longus tendon

Flexor hallucis longus tendon

Flexor digitorum brevis tendon

Flexor digitorum brevis

Abductor digiti minimi

Plantar aponeurosis (cut)

Tuberosity of calcaneous

Lumbricals

Flexor digiti minimi brevis

Quadratus plantae

Flexor hallucis longus tendon

Flexor digitorum longus tendon

Flexor hallucis longus tendon

Tendon of flexor hallucis longus

Sesamoid bones

Adductor hallucis:

Transverse head

Oblique head

Flexor hallucis brevis

Tibialis posterior tendon

Flexor digitorum longus tendon (cut)

Long plantar ligament

Flexor digiti minimi brevis

Plantar interosseus

Peroneus brevis tendon

Peroneus longus tendon

SECTION 13

Foot and Ankle

Figure 13.13 Ligaments and tendons of the foot.

Table 13.5 Muscle Action and Innervation

Action	Muscle involved	Nerve supply	Nerve root
Plantarflexion	Gastrocnemius	Tibial	S1-S2
	Soleus	Tibial	S1-S2
	Plantaris	Tibial	S1-S2
	Flexor digitorum longus	Tibial	S2-S3
	Peroneus longus	Superficial peroneal	L5, S1-S2
	Peroneus brevis	Superficial peroneal	L5, S1-S2
	Flexor hallucis longus	Tibial	S2-S3
	Tibialis posterior	Tibial	L4-L5
Dorsiflexion	Tibialis anterior	Deep peroneal	L4-L5
	Extensor digitorum longus	Deep peroneal	L5, S1
	Extensor hallucis longus	Deep peroneal	L5, S1
	Peroneus tertius	Deep peroneal	L5, S1
Inversion	Tibialis posterior	Tibial	L4-L5
	Flexor digitorum longus	Tibial	S2-S3
	Flexor hallucis longus	Tibial	S2-S3
	Tibialis anterior	Deep peroneal	L4-L5
	Extensor hallucis longus	Deep peroneal	L5, S1
Eversion	Peroneus longus	Superficial peroneal	L5, S1-S2
	Peroneus brevis	Superficial peroneal	L5, S1-S2
	Peroneus tertius	Deep peroneal	L5, S1
	Extensor digitorum longus	Deep peroneal	L5, S1
Flexion of the toes	Flexor digitorum longus	Tibial	S2-S3
	Flexor hallucis longus	Tibial	S2-S3
	Flexor digitorum brevis	Tibial	S2-S3
	Flexor hallucis brevis	Tibial	S2-S3
	Interossei	Tibial	S2-S3
	Flexor digiti minimi brevis	Tibial	S2-S3
	Lumbricalis (MCP)	Tibial	S2-S3
Extension of the toes	Extensor digitorum longus	Deep peroneal	L5, S1
	Extensor hallucis longus	Deep peroneal	L5, S1
	Extensor digitorum brevis	Deep peroneal	S1-S2
	Lumbricales	Tibial	S2-S3
Abduction of the toes	Abductor hallucis	Tibial	S2-S3
	Abductor digiti minimi	Tibial	S2-S3
	Dorsal interossei	Tibial	S2-S3
Adduction of the toes	Adductor hallucis	Tibial	S2-S3
	Plantar interossei	Tibial	S2-S3

DIFFERENTIAL DIAGNOSIS

Table 13.6 Differential Diagnosis

Disorder	Description	Onset	Symptoms	Signs	Special tests
Achilles tendinitis	Tendinitis of Achilles tendon	Overuse	Pain in Achilles tendon with walking and jumping	Swelling in the posterior calcaneus, crepitus in Achilles tendon	
Achilles tendon rupture	Rupture of Achilles tendon	Sudden explosive push-off; males 30-40 yr old	Sudden pain in Achilles tendon region	Inability to walk, incongruency in Achilles tendon	Thompson's test
Ankle sprain	Inversion or eversion twist to ankle causing ligamentous tearing	Sudden turning of the ankle	Immediate pain under lateral or medial malleolus, unable to bear weight	Swelling, discoloration, limp	Anterior drawer, talar tilt
Bursitis	Inflammation of Achilles bursa	Poorly fitting shoes, overuse	Pain in posterior aspect of heel between tendon and skin	Pain with pinching of Achilles bursa, pain with walking	
Hallux rigidus	Limited motion of 1st metatarsal	Chronic, instability of great toe	Pain at base of great toe, aggravated with walking	Decreased mobility of 1st metatarsal	
Metatarsalgia	Pain along the metatarsal with numbness	Overuse	Pain along the metatarsal shaft, associated numbness		
Morton's neuroma	Inflammation of the heads of the 2nd and 3rd metatarsals	Chronic	Pain at the base of metatarsal, usually 2nd and 3rd	Tissue thickening at the area of the MT heads	Tinel's
Plantar fascitis	Inflammation of the plantar fascia	Overuse, acute from stepping on hard object	Heel pain with weight bearing, worse in the morning	Point tender over plantar surface of calcaneus	X-ray may show bone spur

(continued)

SECTION
13

Foot and Ankle

Table 13.6 *(continued)*

Disorder	Description	Onset	Symptoms	Signs	Special tests
Gouty arthritis	Connective tissue disease associated with inflammation of the great toe	Insidious	Pain in the great toe	Swelling in great toe, deformity	X-ray, uric acid test
Hallux valgus	Lateral deviation of the 1st metatarsal	Chronic	May have pain in the great toe	Deformity, bunion formation	X-ray

FUNCTIONAL TESTING OF THE LOWER EXTREMITY

This section begins by describing muscle imbalances associated with the lower extremity. It then lists functional ranges of motion needed for most activities of daily living. The remainder of the section is a list of common functional tests used by clinicians to assess the hip, knee, and ankle.

MUSCLE IMBALANCES

- Tight hip extensors cause increased lumbar flexion when thigh is flexed.
- Tight hip flexors cause increased lumbar extension and anterior tilt as the thigh extends.
- Tight adductors cause lateral pelvic tilt opposite and side-bending toward the side of tightness during weight bearing.
- Tight abductors cause lateral pelvic tilt toward and side-bending away from the side of tightness during weight bearing.

- Short leg results in abduction of the hip on the short side, adduction of hip on long side, convexity of lumbar spine toward short side, iliotibial band (ITB) tightness on short side, and gluteus medius weakness on long side.
- Overuse of two-joint hip flexor muscles rather than iliopsoas may cause faulty hip mechanics or knee pain.
- Overuse of hamstring muscles rather than gluteus maximus may cause overuse problems of the hamstrings.
- Anterior pelvic tilt results in hip flexion and lumbar extension caused by combined shortness of hip flexors and back extensors. Abdominals, glutei, and hamstrings are lengthened.
- Posterior pelvic tilt results in hip extension and lumbar flexion caused by shortness of hip extensors and trunk flexors. Hip flexors and erector spinae are lengthened.

- In forward bending as the head and upper trunk begin flexion, the pelvis moves posteriorly no more than 2 in. to maintain the center of gravity. Continued forward bending is accompanied by hip flexion until approximately 45°. Hip flexion at end-range forward bending should be between 70° and 95°. The return to upright position begins with posterior translation of the pelvis then back extension.

- Pelvic rotation in transverse plane: Rotation occurs around one lower extremity that is fixed on the ground. With weight bearing, forward rotation of the pelvis results in internal rotation of the hip on the stance side. Backward rotation of the pelvis results in external rotation of the hip on the stance side. On forward rotation the adductors, obliques, and sartorius become tight while the gluteus medius and tensor fascia lata become lengthened. On backward rotation, the gluteus medius and tensor become short and the adductors, obliques, and sartorius become lengthened.

- Pelvic elevation in frontal plane: Asymmetry of the pelvis in the frontal plane due to a leg-length difference causes changes in muscle length for many hip muscles. On the long side, the hip is in relative adduction resulting in shortness of the adductors and quadratus lumborum and lengthening of the gluteus medius and tensor fascia lata.

Functional Ranges of Motion Necessary for Lower Extremity

	Gait	Stair ascent	Stair descent
Hip	0-40°	0-60°	0-45°
Knee	0-60°	0-90°	0-100°
Ankle			
Dorsiflexion	0-10°	0-20°	0-20°
Plantarflexion	0-30°	0-30°	0-30°

Adapted from ■ Livingstone, ■ Rancho, ■ McFadyen, and ■ Winter, 1988.

FUNCTIONAL STRENGTH TESTING OF THE HIP

Table 14.1 Functional Tests

Start position	Action	Functional test
Standing	Lift foot onto 20-cm step and return (hip flexion/extension)	5 to 6 reps: functional 3 to 4 reps: functionally fair 1 to 2 reps: functionally poor 0 reps: nonfunctional
Standing	Sit in chair and return to standing (hip flexion/extension)	5 to 6 reps: functional 3 to 4 reps: functionally fair 1 to 2 reps: functionally poor 0 reps: nonfunctional
Standing	Lift leg to balance on one leg keeping pelvis straight (hip abduction)	Hold 1 to 1.5 min: functional Hold 30 to 59 s: functionally fair Hold 1 to 29 s: functionally poor Cannot hold: nonfunctional
Standing	Walk sideways 6 m (hip adduction)	6 to 8 m one way: functional 3 to 6 m one way: functionally fair 1 to 3 m one way: functionally poor 0 m: nonfunctional

Table 14.1   *(continued)*

Start position	Action	Functional test
Standing	Test leg off floor (patient may hold onto something for balance) medially rotate	10 to 12 reps: functional 5 to 9 reps: functionally fair 1 to 4 reps: functionally poor 0 reps: nonfunctional
Standing	Test leg off floor (patient may hold onto something for balance) laterally rotate	10 to 12 repetitions: functional 5 to 9 repetitions: functional fair 1 to 4 repetitions: functional poor 0 repetitions: nonfunctional

Table 14.2   Harris Hip Function Scale

(Circle one in each group)

Pain (44 points)

None/ignores	44
Slight, occasional, no compromise in activity	40
Mild, no effect on ordinary activity, pain after unusual activity, uses aspirin	30
Moderate, tolerable, makes concessions, strong meds	20
Marked, serious limitation	10
Totally disabled	0

Range of Motion (5 points max)

Instructions: Record 10° of fixed adduction as "−10° abduction, adduction to 10°."
Record 10° of fixed external rotation as "−10° internal rotation, external rotation to 10°."
Record 10° of fixed external rotation with 10° further external rotation as "−10° internal rotation, external rotation to 20°."

Function (47 points)

Gait (walking max distance; 33 points max) (1)_____

		Permanent flexion	Range _____°	Index factor	Index value*
1. *Limp*		A. Flexion to			
None	11	(0-45°)		1.0	
Slight	8	(45-90°)		0.6	
Moderate	5	(90-120°)		0.3	
Unable to walk	0	(120-140°)		0.0	
2. *Support*			_____°		
None	11	B. Abduction to			
Cane, long walks	7	(0-15°)		0.8	
Cane, full time	5	(15-30°)		0.3	
Crutch	4	(30-60°)		0.0	
Two canes	2		_____°		
Two crutches	0	C. Adduction to			
Unable to walk	0	(0-15°)		0.2	
3. *Distance walked*		(15-60°)		0.0	
Unlimited	11		_____°		
Six blocks	8	D. External			
Two to three blocks	5	rotation in			
Indoors only	2	extension to			
Bed and chair	0	(0-30°)		0.4	
		(30-60°)		0.0	

(continued)

SECTION 14

Lower Extremity

Table 14.2 *(continued)*

Function (47 points) *(continued)*		Permanent flexion *(continued)*	Range _____°	Index factor	Index value*
Functional activities (14 points max)					
1. *Stairs*		E. Internal rotation in extension to (0-60°)		0.0	
Normally	4				
Normally with banister	2				
Any method	1				
Not able	0				
2. *Socks and tie shoes*		*Index value = range x index factor			
With ease	4				
With difficulty	2	Total index value (A+B+C+D+E) _____			
Unable	0				
3. *Sitting*		Total range of motion points _____			
Any chair, 1 hour	5	(multiply total index value x 0.05)			
High chair, 1/2 hour	3				
Unable to sit 1/2 hour any chair	0	Pain points _____			
4. *Enter public transport*		Function points _____			
Able to use public transport	1	Absence of deformity points _____			
Not able to use public transport	0	Range of motion points _____			
		Total points (100 points max) _____			
Absence of deformity (requires all four; 4 points max)					
1. Fixed adduction ≤ 10°	4	Comments:			
2. Fixed internal rotation in extension < 10°	0				
3. Leg-length discrepancy less than 1-1/4 in.					
4. Pelvic flexion contracture < 30°					

FUNCTIONAL STRENGTH TESTING OF THE KNEE

Table 14.3 Cincinnati Rating Scale

Left	Right	Points	Symptoms (50 points)
			1. Pain
		20	No pain, normal knee, performs 100%.
		16	Occasional pain with strenuous sports or heavy work, knee not entirely normal, some limitations but minor and tolerable.
		12	Occasional pain with light recreational sports or moderate work activities, often brought on by vigorous activities, running, heavy labor, strenuous sports.
		8	Pain, usually brought on by sports, light recreational activities, or moderate work. Occasionally occurs with walking, standing, or light work.
		4	Pain is a significant problem with activities as simple as walking. Relieved by rest. Unable to do sports.

Table 14.3 *(continued)*

Left	Right	Points	
			1. Pain *(continued)*
		0	Pain present all the time, occurs with walking, standing, and at nighttime. Not relieved with rest. I don't know what my pain level is. I have not tested my knee.
			Intensity of pain ☐ Mild ☐ Moderate ☐ Severe **Frequency of pain** ☐ Intermittent ☐ Constant **Location of pain** ☐ Medial ☐ Anterior patella ☐ Posterior diffuse **Pain occurs on** ☐ Stairs ☐ Sitting ☐ Kneeling ☐ Standing **Type of pain** ☐ Sharp ☐ Aching ☐ Throbbing ☐ Burning
		10	**2. Swelling** No swelling, normal knee, 100% activity.
		8	Occasional swelling with strenuous sports or heavy work. Some limitations but minor and tolerable.
		6	Occasional swelling with light recreational sports or moderate work activities, frequently brought on by vigorous activities, running, heavy labor, strenuous sports.
		4	Swelling limits sports and moderate work. Occurs infrequently with simple walking activities or light work (about 3 times/year).
		2	Swelling brought on by simple walking activities and light work. Relieved with rest.
		0	Severe problem all of the time with simple walking activities. I don't know what my swelling level is. I have not tested my knee.
			If swelling occurs it is *(check one box on each line):* **Intensity** ☐ Mild ☐ Moderate ☐ Severe **Frequency** ☐ Intermittent ☐ Constant
		20	**3. Giving way** No giving way.
		16	Occasional giving way with strenuous sports or heavy work. Can participate in all sports but some guarding or limitations are still present.

(contintued)

SECTION
14

Lower Extremity

Table 14.3 *(continued)*

Left	Right	Points	
			3. Giving way (continued)
		12	Occasional giving way with light recreational activities or moderate work. Able to compensate; limits vigorous activities, sports, or heavy work; not able to cut or twist suddenly.
		8	Giving way limits sports and moderate work; occurs infrequently with walking or light work (about 3 times/year).
		4	Giving way with simple walking activities and light work. Occurs once a moth. Requires guarding.
		0	Severe problem with simple walking activities; cannot turn or twist while walking without giving way
			I do not know my level of giving way. I have not tested my knee.
			4. Other symptoms (unscored)

Knee stiffness
- ☐ None
- ☐ Occasional
- ☐ Frequent

Kneecap grinding
- ☐ None
- ☐ Mild
- ☐ Moderate
- ☐ Severe

Knee locking
- ☐ None
- ☐ Occasional
- ☐ Frequent

Left	Right	Points	
			Function (50 points)
			5. Overall activity level
		20	No limitation, normal knee, able to do everything, including strenuous sports or heavy labor.
		16	Perform sports, including vigorous activities, but at a lower performance level; involves guarding or some limits to heavy labor.
		12	Light recreational activities possible with rare symptoms; more strenuous activities cause problems.
		8	No sports or recreational activities possible. Walking activities possible with rare symptoms; limited to light work.
		4	Walking, activities of daily living cause moderate symptoms, frequent limitations.
		0	Walking, activities of daily living cause severe problems, persistent symptoms.
			I do not know what my real activity level is, I have not tested my knee, or I have given up strenuous sports.
			6. Walking
		10	Normal, unlimited.
		8	Slight/mild problem.
		6	Moderate problem: smooth surface possible up to 800 m.
		4	Severe problem: only 2-3 blocks possible.
		2	Severe problem: requires cane, crutches.

Table 14.3 *(continued)*

Left	Right	Points	
			7. Stairs
		10	Normal, unlimited.
		8	Slight/mild problem.
		6	Moderate problem: only 10-15 steps possible.
		4	Severe problem: requires banister support.
		2	Severe problem: only 1-5 steps possible.
			8. Running activity
		5	Normal, unlimited: fully competitive, strenuous.
		4	Slight/mild problem: run half speed.
		3	Moderate problem: only 2-4 km possible.
		2	Severe problem: only 1-2 blocks possible.
		1	Severe problem: only a few steps.
			9. Jumping or twisting activities
		5	Normal, unlimited: fully competitive strenuous.
		4	Slight/mild problem: some guarding, but sports possible.
		3	Moderate problem: gave up strenuous sports, recreational sports possible.
		2	Severe problem: affects all sports; must constantly guard.
		1	Severe problem: only light activity possible (golf, swimming).

TOTAL: Left _____ Right _____ (Maximum: 100 points)

Table 14.4a Lysholm Knee Scale

Limp (5 points)			**One step at a time**	2	_____
None	5	_____	Unable	0	_____
Slight or periodic	3	_____			
Severe and constant	0	_____	**Squatting (5 points)**		
			No problems	5	_____
Support (5 points)			Slightly impaired	3	_____
Full support	5	_____	Not past 90 degrees	2	_____
Cane or crutch	3	_____	unable	0	_____
Weight bearing impossible	0	_____			
Stair climbing (10 points)			**Total**		_____
No problems	5	_____			
Slightly impaired	3	_____			

SECTION
14

Lower Extremity

Table 14.4b   Walking, Running, Jumping

Instability		Pain	
Never giving way	30 _____	None	30 _____
Rarely gives way except for athletic or other severe exertion	25 _____	Inconstant and slight during severe exertion	25 _____
Gives way frequently during athletic events or other severe exertion	0 _____	Marked on giving way	20 _____
		Marked during severe exertion	15 _____
Occasionally in daily activities	10 _____	Marked on or after walking more than 1-1/4 miles	10 _____
Often in daily activities	5 _____	Marked on or after walking less than 1-1/4 miles	5 _____
Every step	0 _____	Constant and severe	0 _____
Swelling		**Atrophy of thigh (5 points)**	
None	10 _____	None	5 _____
With giving way	7 _____	1-2 cm	3 _____
On severe exertion	5 _____	> 2 cm	0 _____
On ordinary exertion	2 _____		
Constant	0 _____		
		Total	_____

FUNCTIONAL STRENGTH TESTING OF THE FOOT AND ANKLE

Table 14.5   Functional Tests

Start position	Action	Functional test
Standing on one leg	Lift toes and forefeet off ground	10 to 15 reps: functional 5 to 9 reps: functionally fair 1 to 4 reps: functionally poor 0 reps: nonfunctional
Standing on one leg	Lift heels off ground	10 to 15 reps: functional 5 to 9 reps: functionally fair 1 to 4 reps: functionally poor 0 reps: nonfunctional
Standing on one leg	Lift lateral aspect of foot off ground	10 to 15 reps: functional 5 to 9 reps: functionally fair 1 to 4 reps: functionally poor 0 reps: nonfunctional
Standing on one leg	Lift medial aspect of foot off ground	10 to 15 reps: functional 5 to 9 reps: functionally fair 1 to 4 reps: functionally poor 0 reps: nonfunctional

Table 14.5 *(continued)*

Start position	Action	Functional test
Seated	Pick up and released marbles	10 to 15 reps: functional 5 to 9 reps: functionally fair 1 to 4 reps: functionally poor 0 reps: nonfunctional
Seated	Lift toes off ground	10 to 15 reps: functional 5 to 9 reps: functionally fair 1 to 4 reps: functionally poor 0 reps: nonfunctional

Table 14.6 Functional Rating Scale for Ankles

Condition	Score		Condition	Score
Pain Never hurts Hurts with strenuous sports Hurts with light sports Hurts with walking more than 5 km Hurts walking less than 5 km Hurts at rest or at night	5 4 3 2 1 0		**Swelling** Never swells End of day or after activity Swells most of time	2 1 0
Stability Never turns (no support) Never turns (with support) Turns occasionally (no support) Turns frequently during daily living activities	5 4 3 0		**Activity** Able to hop Not as good as normal side Cannot hop	2 1 0
Stiffness Never stiff Stiff untill warmed up Stiff at all times	2 1 0		**Range of motion** Full range of motion Full plantarflexion, limited dorsiflexion Limited dorsi- and planterflexion	4 2 0
			Total	

OTHER TESTS

Table 14.7 Physical Performance Test

Task	Performance **	Scoring		Score
1. Lift a book and put it on a shelf	_____ sec	2 sec 2.5-4 sec 4.5-6 sec > 6 sec Unable	= 4 = 3 = 2 = 1 = 0	_____
2. Put on and remove a jacket or sweatshirt	_____ sec	10 sec 10.5-15 sec 15.5-20 sec > 20 sec Unable	= 4 = 3 = 2 = 1 = 0	_____
3. Pick up a penny/pencil from the floor	_____ sec	2 sec 2.5-4 sec 4.5-6 sec > 6 sec Unable	= 4 = 3 = 2 = 1 = 0	_____
4. Turn 360 degrees	Manner of steps	Discontinuous steps = 0 Continuous steps = 2 Unsteady (staggers) = 0 Steady = 2		_____ _____
5. 50 ft. walk test	_____ sec	15 sec 15.5-20 sec 20.5-25 sec > 25 sec Unable	= 4 = 3 = 2 = 1 = 0	_____
6. Climb one flight of stairs*	_____ sec	5 sec 5.5-10 sec 10.5-15 sec >15 sec Unable	= 4 = 3 = 2 = 1 = 0	_____
7. Climb stairs*	Number of flights of stairs (9-12 steps) up and down (maximum 4)	1 flight 2 flights 3 flights 4 flights	= 1 = 2 = 3 = 4	_____

Total score (maximum 28 for 7 items, 20 for 5 items) _____ 7-item _____ 5-item

*For timed measurements, round to the nearest 0.5 seconds.

**May omit #6 and #7 if flight of stairs is not available.

Table 14.8 Physical Scales of the AIM 2

Please check (X) the most appropiate answer for each question.	All days (1)	Most days (2)	Some days (3)	Few days (4)	No days (5)
These questions refer to mobility level.					
DURING THE PAST MONTH					
1. How often were you physically able to drive a car or use public transportation?	____	____	____	____	____
2. How often were you out of the house for at least part of the day?	____	____	____	____	____
3. How often were you able to do errands in the neighborhood?	____	____	____	____	____
4. How often did someone have to assist you to get around outside your home?	____	____	____	____	____
5. How often were you in a bed or chair for most or all of the day?	____	____	____	____	____
These questions refer to walking and bending.					
DURING THE PAST MONTH					
6. Did you have trouble doing vigorous activities such as running, lifting heavy objects, or participating in strenuous sports?	____	____	____	____	____
7. Did you have trouble either walking several blocks or climbing a flight of stairs?	____	____	____	____	____
8. Did you have trouble bending, lifting, or stooping?	____	____	____	____	____
9. Did you have trouble either walking one block or climbing a flight of stairs?	____	____	____	____	____
10. Were you unable to walk unless assisted by another person or by a cane, crutches, or walker?	____	____	____	____	____
These questions refer to hand and finger function.					
DURING THE PAST MONTH					
11. Could you easily write with a pen or pencil?	____	____	____	____	____
12. Could you easily button a shirt or blouse?	____	____	____	____	____
13. Could you easily turn a key in a lock?	____	____	____	____	____
14. Could you easily tie a knot or a bow?	____	____	____	____	____
15. Could you easily open a new jar of food?	____	____	____	____	____

(continued)

Table 14.8 *(continued)*

	All days (1)	Most days (2)	Some days (3)	Few days (4)	No days (5)
These questions refer to mobility level.					
DURING THE PAST MONTH					
16. Could you easily wipe your mouth with a napkin?	_____	_____	_____	_____	_____
17. Could you easily put on a pullover sweater?	_____	_____	_____	_____	_____
18. Could you easily comb or brush your hair?	_____	_____	_____	_____	_____
19. Could you easily scratch your lower back with your hand?	_____	_____	_____	_____	_____
20. Could you easily reach shelves that were above your head?	_____	_____	_____	_____	_____
These questions refer to self-care tasks.					
DURING THE PAST MONTH					
21. Did you need help to take a bath or shower?	_____	_____	_____	_____	_____
22. Did you need help to get dressed?	_____	_____	_____	_____	_____
23. Did you need help to use the toilet?	_____	_____	_____	_____	_____
24. Did you need help to get in or out of bed?	_____	_____	_____	_____	_____
These questions refer to household tasks.					
DURING THE PAST MONTH					
25. If you had the necessary transportation, could you go shopping for groceries without help?	_____	_____	_____	_____	_____
26. If you had kitchen facilities, could you prepare your own meals without help?	_____	_____	_____	_____	_____
27. If you had household tools and appliances, could you do your own housework without help?	_____	_____	_____	_____	_____
28. If you had laundry facilities, could you do your laundry without help?	_____	_____	_____	_____	_____

Table 14.9 Tinetti Balance and Gait Tests

Tinetti Balance Test	
Subject is seated in hard, armless chair. The following maneuvers are tested.	
1. Sitting balance	
Leans or slides in chair	= 0
Steady, safe	= 1 _____
2. Arises	
Unable without help	= 0
Able, uses arms to help	= 1
Able, without using arms	= 2 _____
3. Attempts to arise	
Unable without help	= 0
Able, requires > 1 attempt	= 1
Able to arise, 1 attempt	= 2 _____
4. Immediate standing balance (first five seconds)	
Unsteady (swaggers, moves feet, trunk sways)	= 0
Steady but uses cane or other support	= 1
Steady without walker or other support	= 2 _____
5. Standing balance	
Unsteady	= 0
Steady but wide stance (medial heels > 4 in. apart) and uses cane or other support	= 1
Narrow stance without support	= 2 _____
6. Nudged (subject at maximum position with feet as close together as possible; examiner pushes lightly on subject's sternum with palm of hand 3 times)	
Begins to fall	= 0
Staggers, grabs, catches self	= 1
Steady	= 2 _____
7. Eyes closed (at maximum position no. 6)	
Unsteady	= 0
Steady	= 1 _____
8. Turning 360 degrees	
Discontinuous steps	= 0
Continuous	= 1
Unsteady (grabs, staggers)	= 0
Steady	= 1 _____

(continued)

SECTION
14

Lower Extremity

Table 14.9 *(continued)*

Tinetti Balance Test	
9. Sitting down	
Unsafe (misjudged distance, falls into chair) Uses arms or not a smooth motion Safe, smooth motion	= 0 = 1 = 2 _____
Balance score: **/16**	

Tinetti Gait Test	
Subject stands with examiner, walks down hallway or across room, first at "usual" pace, then back at "rapid" pace using usual walking aids.	
10. Initiation of gait (immediately after told to "go")	
Any hesitancy or multiple attempts to start No hesitancy	= 0 = 1 _____
11. Step length and height **a. Right swing foot**	
Does not pass left stance foot with step Passes left stance foot Right foot does not clear floor completely with step Right foot completely clears floor	= 0 = 1 = 0 = 1 _____
b. Left swing foot	
Does not pass right stance foot with step Passes right stance foot Left foot does not clear floor completely with step Left foot completely clears floor	= 0 = 1 = 0 = 1 _____
12. Step symmetry	
Right and left step length not equal (estimate) Right and left step appear equal	= 0 = 1 _____
13. Step continuity	
Stopping or discontinuity between steps Steps appear continuous	= 0 = 1 _____
14. Path (estimated in relation to floor tiles, 12 in. diameter; observe excursion of one foot over about 10 ft of the course)	
Marked deviation Mild/moderate deviation or uses walking aid Straight without walking aid	= 0 = 1 = 2 _____

Table 14.9 *(continued)*

Tinetti Gait Test	
15. Trunk	
Marked sway or uses walking aid	= 0
No sway but flexion of knees or back or spreads arms out while walking	= 1
No sway, no flexion, no use of arms, and no use of walking aid	= 2 _____
16. Walking time	
Heels apart	= 0
Heels almost touching while walking	= 1 _____

Gait score: /12
Balance + gait score: /28

PART

V

POSTURE
AND GAIT

SECTION 15

POSTURE

This section reviews normal and abnormal standing posture in the posterior and lateral views. Normal posture is probably a misnomer in that most people do not possess "normal" posture. However, by identifying normal or ideal posture the clinician has a posture with which to make comparisons. This section identifies muscle imbalances associated with faulty posture.

TERMINOLOGY

Lordosis: anterior curve of the spine

Kyphosis: posterior curve of the spine

Scoliosis: lateral curvature of the spine

Cervical curvature: convex anteriorly

Thoracic curvature: convex posteriorly

Lumbar curvature: convex anteriorly

Sacral curvature: convex posteriorly

IDEAL POSTURE

Below are some standards of "normal" posture against which to compare imbalances. Plumb line alignment, distribution of weight, and electromyography (EMG) activity are three means of assessing posture.

Plumb Line Alignment

1. Slightly anterior to the lateral malleolus; through calcaneocuboid joint

2. Just in front of the center of the knee joint

3. Through the greater trochanter of femur; slightly posterior to center of hip joint

4. Midway through trunk; through bodies of lumbar vertebrae

5. Through shoulder joint

6. Through bodies of cervical vertebrae

7. Through odontoid process

8. Through lobe of the ear; through external auditory meatus

Pressure Distribution of Weight

45-65% of body weight should be carried over heels

30-47% of body weight should be carried on the forefoot

1-8% of body weight should be carried over the midfoot

EMG Activity in Quiet Standing

1. Muscles of the feet are quiet.

2. Soleus is active to maintain upright position.

3. Quadriceps and hamstrings for the most part are quiescent, although they may show slight activity from time to time.

4. Iliopsoas remains constantly active.

5. Gluteus maximus is quiescent.

215

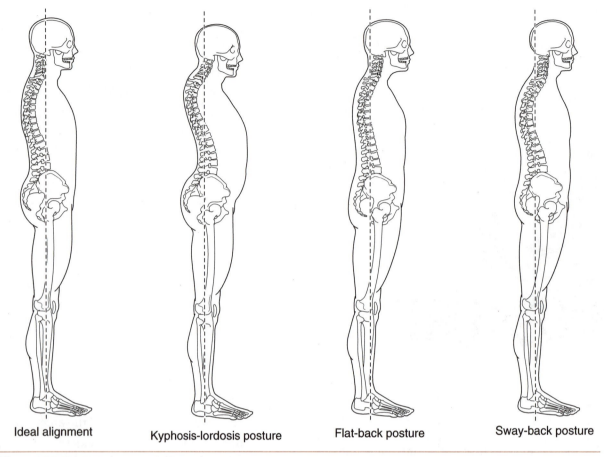

Ideal alignment   Kyphosis-lordosis posture   Flat-back posture   Sway-back posture

Figure 15.1   Standing postures.

6. Gluteus medius and tensor fascia lata are active to control lateral pelvic tilt.

7. Erector spinae is active counteracting anterior moment.

8. Abdominal muscles remain quiescent.

9. Minimal activity in the upper trapezius, serratus anterior, supraspinatus, posterior deltoid.

Table 15.1   Body Posture

Good posture	Part	Faulty posture
In standing, the longitudinal arch has the shape of a half-dome. The feet toe out slightly.	Foot	Low longitudinal arch or flat feet. Low metatarsal arch, usually indicated by calluses under the foot. Weight borne on inner or outer border of the foot. Toeing out or toeing in.
Toes should be straight without curling downward or bent upward. They should extend forward in line with the foot.	Toes	Hallux valgus, or curling of the toes as in hammertoes.
Legs are straight in frontal and sagittal planes. Patellae face straight ahead.	Knee and legs	Genu valgus, genu varum, genu recurvatum, flexed knee. Patellae ride high/low/inward/outward.
Body weight distributed evenly, hips are symmetrical. Spine does not curve (slight deviation to the left for right-handed and vice-versa for left-handed is common).	Hips, pelvis, spine (back view)	One hip is higher than the other. Hips are rotated so one is anteriorly rotated.
Front of the pelvis and thighs are in straight line. ASIS and pubic physis aligned. Four natural curves are present; in the neck and low back the curve is forward; in the upper back and sacral region it is backward.	Hips and pelvis (side view)	Increased lordosis, anterior pelvic tilt. Low back is flat, pelvis in sym-posterior tilt. Kyphosis in the thoracic spine. Increased forward head.
In adults the abdomen should be flat.	Abdomen	Abdomen protrudes.
Chest is slightly up and slightly forward.	Chest	Hollow chest; lifted and held too high by arching in the back; ribs are not symmetrical.
Arms hang relaxed and equidistant from sides with palm of the hands facing toward the body. Shoulders are level in all planes. Scapulae lie flat against the rib cage. There is about a 4 in. separation between them.	Arms and shoulders	Shoulders are forward and are assymmetrical. One arm hangs lower than the other. Scapulae abducted, winged, or excessively rotated.
Head is held erect in a position of good balance.	Head	Chin up too high. Head protrudes forward. Head tilted or rotated.
Mouth remains closed.	Jaw	Mouth hangs open.

Table 15.2 Postural Musculature

Postural fault	Anatomic position	Muscles (short)	Muscles (long)
Lordosis	Lumbar spine hypererextension Anterior tilt pelvis Hip joint flexion	Lower back erector spinae Hip flexors	Abdominals External obliques Hip extensors
Flat back	Lumbar spine flexion Posterior tilt pelvis Hip joint extension	Anterior abdominals Hip extensors	Erector spinae Hip flexors
Sway-back	Posterior tilt Hip joint extension	Rectus abdominus Internal obliques Hip extensors	External obliques One joint Hip flexors
Forward head	Cervical spine hyperextension	Cervical spine extensors Upper trapezius, levator	Cervical spine flexors
Kyphosis	Thoracic spine flexion ↓ intercostal space	Internal obliques (upper lateral) Shoulder adductors Pectoralis minor intercostals	Thoracic spine extensors Mid trapezius Low trapezius
Forward Shoulders	Scapuae abductors Scapulae elevated	Serratus anterior Pectoralis minor and upper trapezius	Mid trapezius Low trapezius
Scoliosis Slight left C-curve Thoracolumbar scoliosis	Thoracolumbar spine lateral flexion Convex toward left	Right latissimus dorsi Trunk muscles Left psoas	Lattissimus dorsi Trunk muscles Right psoas
(Opposite for right C-curve)			
High Right hip	Pelvis, lateral tilt high on right Right hip adduction Left hip abduction	Right lateral trunk Left hip abduction, TFL Right hip adduction	Left lateral trunk Right hip abduction, gluteus medius Left hip adduction
(Opposite for posture with right C-curve and high left hip)			
Hyperextended knee	Knee hyperextension Ankle plantar flexion	Quadriceps Soleus	Popliteus Hamstrings

Table 15.2 *(continued)*

Postural fault	Anatomic position	Muscles (short)	Muscles (long)
Flexed knee	Knee flexion	Popliteus Hamstrings	Quadriceps Soleus
Medial rotation of femur	Hip joint medial rotation	Hip medial rotators	Hip lateral rotators
Genu valgum	Hip joint adduction Knee joint abduction	Fascia lata Lateral knee joint structures	Medial knee joint structures
Postural bowlegs	Hip joint medial rotators Knee joint hyperextension Foot pronation	Hip medial rotators Quadriceps, foot everters	Hip lateral rotators Popliteus; posterior tibialis Long toe flexors
Pronation	Calcaneal eversion	Peroneals Toe extensor	Post tibialis Long toe flexor
Supination	Calcaneal inversion	Tibials	Peroneals
Hammertoes Low MT arch	MTP extension PIP flexion	Toe extensors	Lumbricales

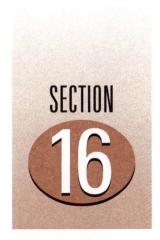

SECTION 16

GAIT

Gait assessment should be a component of every orthopedic evaluation. This section facilitates gait evaluation by offering information in an accessible manner. The section lists common terminology used with gait evaluation, describes the gait sequence, and includes a table that lists joint motion and muscle activity associated with each phase of gait. The section also identifies common faulty gait characteristics (figure 16.1).

TERMINOLOGY

Gait cycle: from heel strike to heel strike of the same foot.

Stance phase: phase of gait in which one foot contacts the ground and remains in contact with the ground; includes initial contact, loading response, midstance, terminal stance, and pre-swing.

Swing phase: phase of gait that is determined when one foot leaves the ground until that same foot returns to the ground; includes initial swing, midswing, and terminal swing.

Stance time: the amount of time spent in stance; includes single support and double support.

Swing time: the amount of time spent in swing.

Double support: the phase of gait during which both lower extremities are in contact with the ground.

Single support: the phase of gait during which only one lower extremity is in contact with the ground.

Ground reaction force: force created as a result of the foot contact with supporting surface.

Stride length: the distance between two successive events accomplished by same lower extremity.

Stride duration: the amount of time spent in stride.

Step length: the distance between heel strike of one leg and heel strike of the contralateral leg.

Step duration: the time between heel strike of one leg and heel strike of the contralateral leg.

Cadence: number of steps per unit time.

Step width: the linear distance between the midpoint of the heel on one foot and same point on the other foot.

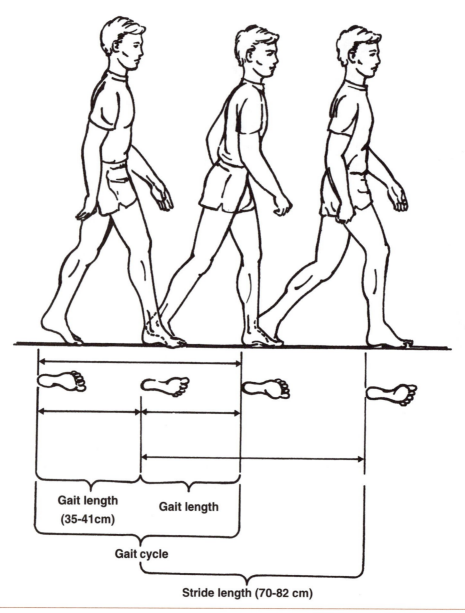

Gait length
(35-41cm)

Gait length

Gait cycle

Stride length (70-82 cm)

Figure 16.1 Gait sequence.

GAIT SEQUENCE

Stance (60% of Gait Cycle)

(Rancho Los Amigos terminology is used; traditional terms are in parentheses).

1. **1% initial contact (heel strike):** contact of the foot with the ground

2. **2-10% loading response (foot flat):** from initial contact until the point at which the contralateral leg leaves the ground (figure 16.2)

3. **10-30% midstance:** from loading response until the body is directly over the supporting limb

4. **30-50% terminal stance (heel-off):** from midstance to a point just prior to initial contact of the contralateral extremity

5. **50-60% pre-swing (toe-off):** from terminal stance to just prior to the liftoff of the reference extremity

Swing (40% of Gait Cycle)

1. **60-73% initial swing (acceleration):** the point of lift to maximum knee flexion

2. **73-87% midswing:** from maximum knee flexion to the point at which the tibia is in a vertical position (figure 16.3)

3. **87-100% terminal swing (deceleration):** from midswing to initial contact

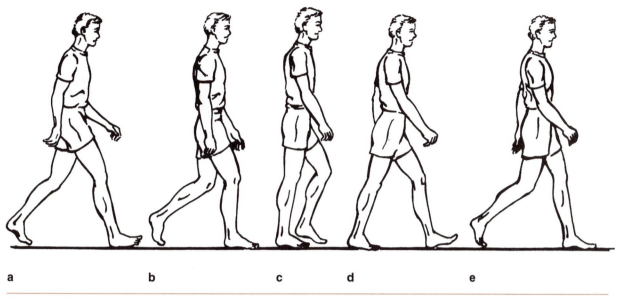

Figure 16.2 Stance phase of gait: (*a*) initial contact, (*b*) loading response, (*c*) midstance (single-leg stance), (*d*) terminal stance, and (*e*) pre-swing.

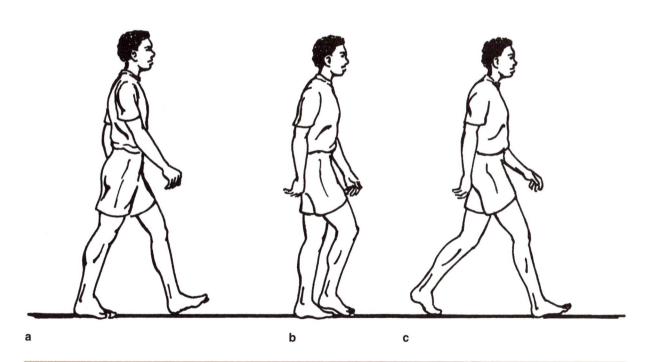

Figure 16.3 Swing phase of gait: (*a*) initial swing (acceleration), (*b*) midswing, and (*c*) terminal swing (deceleration).

GAIT SEQUENCE WITH JOINT POSITION AND MUSCLE ACTIVITY

Table 16.1 Gait With Joint Position and Muscle Activity

Gait phase	Joint and muscle	Pelvis	Hip	Knee	Ankle	Foot
Initial contact	Joint position	Level in sagittal and coronal planes; forward in tranverse plane on stance leg	30° hip flexion, internal rotation	Knee extension (0-5°)	Neutral	2° inverted
	Muscle activity	Gluteus meduis, adductors eccentric	Hamstrings, gluteus maximus eccentric	Vasti, ITB eccentric hamstrings concentric	Pre-tibiales eccentric	Post tibiales eccentric
Loading response	Joint position	Contra-lateral pelvis drops	20° flexion, internal rotation	15° flexion	Neutral to 10° plantar flexion (pf)	Pronated 5°
	Muscle activity	Gluteus medius, minimus, TFL eccentric	Gluteus maximus, adductor magnus concentric	Quadriceps eccentric, slight hamstring	Pre-tibiales eccentric	Post tibiales eccentric
Midstance	Joint position	Level	Neutral, external rotation	0-5° flexion	8° pf to 5° dorsiflexion (df)	Supinated
	Muscle activity	Gluteus medius, TFL concentric	Gluteus medius concentric	Early: quadriceps concentric	Soleus, gastrocnemius eccentric	Post tibiales concentric
Terminal stance	Joint position	Anterior tilt, pelvis rotates 5° backward	Extension 20°, external rotation, abduction	Extension	5° to 10° df	Supinated (maximal) MTP extends to 30°
	Muscle activity	Rectus abdominus	TFL eccentric	Hamstrings eccentric, gastrocnemius	Soleus eccentric	Peroneals eccentric
Pre-swing	Joint position	Lateral tilt ipsilateral side	Neutral to flex, external rotation	40° flexion	20° pf	Supinated MTP to 60°
	Muscle activity	Hip adduction	Add longus eccentric,	Popliteus concentric,	Pf concentric;	Personeals, FDL, EHL

The Clinical Orthopedic Assessment Guide

Table 16.1 *(continued)*

Gait phase	Joint and muscle	Pelvis	Hip	Knee	Ankle	Foot
Pre-swing (continued)		eccentric	iliopsoas concentric; late: hamstrings eccentric	rectus femoris eccentric	pre-tibiales late	concentric
Initial swing	Joint position	Posterior tilt, lateral tilt	Flexion, internal rotation	60° flexion	5° pf to df	Supinated
	Muscle activity	Gluteus medius	Iliopsoas concentric	Biceps femoris concentric	Pre-tibiales concentric	Peroneals concentric
Midswing	Joint position	Posterior tilt	30° flexion, adduction, internal rotation	Flexion to extension	Dorsi-flexion	
	Muscle activity	Gluteus medius	Adductors concentric	Sarterius concentric	Anterior tibiales EHL	
Terminal swing	Joint position	Anteror tilt, rotated forward	30° flexion, internal rotation	Extension	Neutral	
	Muscle activity	Adductors	Gluteus maximus, hamstrings eccentric	Quadriceps concentric	Anterior tibiales isometric, concentric	

RANGE OF MOTION NEEDED DURING GAIT

Stance Phase

Hip flexion: 0-30°

Hip extension: 0-20°

Knee flexion: 0-40°

Knee extension: 0°

Plantarflexion: 0-20°

Dorsiflexion: 0-10°

Subtalar joint inversion: 35°

Subtalar joint eversion: 15°

Swing Phase

Hip flexion: 0-30°

Hip extension: 0-20°

Knee flexion: 0-60°

Knee extension: 0°

Plantarflexion: 0-10°

Dorsiflexion: 0°

GAIT DISTURBANCES AND MECHANICAL FAULTS

Initial Contact

Pelvis: Pelvic drop.

Hip: Excessive hip flexion (greater than 30°); tightness in iliopsoas; limited hip flexion—weakness of hip flexors, gluteus maximus weakness.

Knee: Excessive knee flexion—knee remains in flexion during initial contact; possible causes—weak quadriceps, short leg on contralateral side, joint pain.

Ankle and foot: Foot slap—at heel strike, fore-foot slaps to the ground; possible causes include weak dorsiflexors.

Toes first: Toes contact ground instead of heel; possible causes—leg-length discrepancy, contracted heel cord, plantarflexion contracture, painful heel, flaccidity of dorsiflexors.

Flat footed: Entire foot contacts ground at heel strike—capsular restriction at ankle.

Loading Response

Hip

a. Limited hip extension (does not return to neutral): tight hip flexors

b. Internal rotation: weak external rotators, anteversion of the hip

c. External rotation: retroversion of the hip

Knee: knee hyperextension—weak quadriceps, gastrocnemius.

Ankle: limited plantarflexion—joint restriction, decreased muscle strength.

Midstance

Pelvis

a. Lateral trunk lean: gluteus medius weakness

b. Backward trunk lean: weak gluteus maximus

c. Forward trunk lean: quadriceps weakness

Hip

a. Excessive lordosis: tight hip flexors

b. Contralateral hip drop: weakness of the gluteus medius

c. Increased knee flexion: decreased hip extension, decreased back extension

Knee

a. Knee hyperextension: weak quadriceps, gastrocnemius

b. Patella faces outward: retroversion

c. Knee flexion: capsular restriction of the knee

Ankle and foot

a. Early heel lift: heel lifts early in midstance; tightness in dorsiflexion (soft tissue or joint)

b. Toe clawing: toes grab floor, tightness in intrinsic musculature, poor stability of first metatarsal

c. Valgus deviation: femoral anteversion, overpronation, late pronation

Terminal Swing

Pelvis

a. Medial heel whip: weakness in hip external rotators

b. Hip flexion: hip extension restriction

Knee: excessive knee flexion—more than 40° of knee flexion during push-off.

Ankle and foot: medial heel whip—tight gastrocnemius/soleus complex.

Pre-Swing

Hip: lack of push-off—hip extension restricted.

Ankle and foot: no push-off—insufficient transfer of weight from lateral heel to medial forefoot, limited range of motion; pain in forefoot.

Swing

Hip

a. Circumduction: weak hip flexors, long leg

b. Hip hiking: lack of knee flexion or ankle dorsiflexion, weak hip flexors

c. Excessive hip flexion: weak dorsiflexion

Knee: limited knee flexion—less than 65° of knee flexion, joint pain, limited range of motion.

Ankle and foot

a. Toe drag: insufficient dorsiflexion, weak dorsiflexors and toe extensors

b. Varus: foot excessively inverted—weak peroneals

SECTION

17

OTHER GAIT SEQUENCES

The gait used in running and stair climbing with or without assistive devices do differ from the walking gait. This section highlights these differences.

RUNNING

Running is different from walking in three distinct ways.

Running involves:

1. a double-float period in which both lower extremities are off the ground,
2. occurrence of motions at hips through a greater range except for hip extension, and
3. a ground reaction force that is four to seven times body weight.

The running gait sequence includes the contact and swing phases.

Contact Phase

One foot is in contact with the ground; 38% of running cycle.

Foot strike: Runner contacts the ground with either heel, midfoot, or forefoot.

Midsupport phase: Foot is loose and adapting to terrain, absorbs shock; begins to supinate as heel rises from ground.

Take-off: Foot leaves the ground.

Swing Phase

Leg and foot are off the ground; 62% of running cycle.

Follow-through: End of backward momentum of leg.

Forward swing: Limb begins to drive forward.

Foot descent: Limb prepares for foot strike.

SHOE WEAR

Clients often ask their clinicians what type of footwear is appropriate for them. Different types of feet require different types of shoes (figure 17.1). Rigid feet need shoes with more cushioning and shock absorption, and flexible feet need a more stable shoe. It is also important that shoes fit properly according to the various components of the shoe in the list that follows.

Table 17.1   Running Gait

Gait phase	Joint and muscle	Hip	Knee	Foot
Foot strike	Joint motion	20-50° flexion	15° flexion	10° dorsiflexion
	Muscle activity	Gluteus maximus, gluteus medius, TFL eccentric	Hamstrings, gastrocnemius popliteus concentric; quad. co-contraction	Anterior tibials, toes extensors eccentric
Mid support	Joint motion	30° flexion	20° flexion	20° dorsiflexion
	Muscle activity	Gluteus medius and TFL active control pelvis; gluteus maximus, hamstring eccentrically control limb in flexion	Quadriceps eccentric to control knee flexion	Gastrocsoleus, posterior tibials eccentric
Take-off	Joint motion	10° extension	0° flexion	25° plantar-flexion
	Muscle activity	Hamstrings, gluteus maximus, gluteus medius concentric, trunk musculature eccentric	Quadriceps eccentric	Gastrocsoleus, peroneals, toe flexors concentric; toe extensors eccentric
Follow-through	Joint motion	5° extension	20° flexion	10° plantar-flexion
	Muscle activity	Adductors to control pelvis, hip flexors eccentric to control hip extensors	Medial hamstrings concentric	Gastrocnemius concentric
Forward swing	Joint motion	10-60° flexion	125° flexion	10° plantar-flexion
	Muscle activity	Iliopsoas, rectus femoris, TFL concentric	Hamstrings and quadriceps co-contract	Pre-tibiales eccentrically control ankle
Foot descent	Joint motion	40° flexion	40-20° flexion	10° dorsiflexion
	Muscle activity	Gluteus maximus and hamstrings decelerate flexing thigh, gluteus medius, TFL concentric	Hamstrings eccentric	Pre-tibials concentric

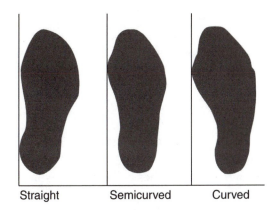

Straight   Semicurved   Curved

Slip lasting

Board lasting

Figure 17.1 Shoe types.

SHOE SHAPE

The bottom of the shoe (last) can be shaped either straight, curved, or semicurved (see figure 17.1). The straight last is for the individual who overpronates, and the curved last is for the individual who has a rigid, high-arched foot.

Midsole

The cushioned layers between the foot and the ground can be made of varying densities of materials. Both the inside and the outside of the shoe may have various densities. Generally the inside should be more firm than the outside.

Heel Counter

The heel counter is the portion of the shoe that stabilizes the rearfoot. A firm heel counter is necessary for all shoes but is more important for individuals with a flexible foot.

Inner Last

The inner last refers to the inner border of the shoe. Underneath the sock liner (insole) on the

inside of the shoe is either stitching (slip lasting) or a piece of fiberboard laminated to the shoe (board lasting). A slip-lasted shoe is used for the rigid foot whereas a board last is used for the more flexible foot. Combination lasts are for combination feet.

STAIR CLIMBING

Ascent (0-14%): Weight Acceptance Through 14-32% (Pull-Up)

Motion

- Hip extension from 60° to 30° of flexion
- Knee extension from 80° to 35° of flexion
- Ankle dorsiflexion from 20° to 25° (figure 17.2)

Muscles Used

Gluteus maximus, semitendinosus, gluteus medius, vastus lateralis, rectus femoris, anterior tibials, soleus, gastrocnemius; all concentric.

32-64%: Forward Continuance (Pull-Up Through Forward Continuance)

Motion

- Hip extension from 30° to 50° flexion
- Hip flexion from 5° to 10° flexion
- Knee extension from 35° to 10° flexion
- Knee flexion from 5° to 10° flexion
- Ankle plantarflexion: 15° dorsiflexion to 10° plantarflexion

Muscles Used

- Gluteus maximus, gluteus medius, semitendinosus—concentric
- Gluteus maximus, gluteus medius—eccentric
- Vastus lateralis, rectus femoris—concentric
- Rectus femoris, vastus lateralis—eccentric
- Soleus—concentric
- Gastrocnemius, anterior tibialis—eccentric

64-100%: Foot Clearance Through Foot Placement

Motion

- Hip flexion from 10° to 60° flexion
- Hip extension from 40° to 50° flexion

SECTION
17

Other Gait

- Knee flexion from 10° to 90° flexion
- Knee extension from 90° to 85° flexion
- Ankle dorsiflexion: 10° of plantarflexion to 20° dorsiflexion

Muscles Used

- Gluteus medius—concentric
- Semitendinosus and semimembranosus—concentric
- Vastus lateralis—concentric
- Rectus femoris, anterior tibialis—concentric
- Leg pull: iliopsoas (concentric), rectus femoris
- Foot placement: hamstrings
- Pull-up: knee extension, quadriceps
- Forward propelling: ankle
- Descent: hamstrings at toe-off
- Leg pull: hip flexors
- Weight acceptance: knee—eccentric
- Support: knee—eccentric

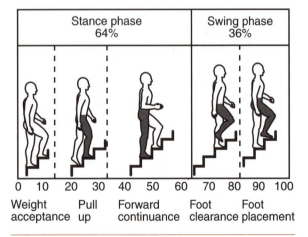

Stair gait cycle
ascent

Figure 17.2 Gait cycle phases during stair climbing.

Assisted Gait

Three-point gait: crutches and affected limb move together, nonweight bearing, modified with toe-touch weight bearing; advance crutches with involved leg.

Four-point gait: one crutch, opposite limb, other crutch, other limb.

Two-point gait: right crutch and left leg together, left crutch and right leg together.

Swing-to gait: crutches and then leg swung forward (paraplegics).

Swing-through gait: crutches and then legs swung forward through the crutches.

Stair Climbing With Assistive Device

Cane

Ascend: Uninvolved lower extremity leads up cane and involved leg.

Descend: Involved limb and cane lead down; uninvolved leg follows.

Crutches (Nonweight Bearing)

Ascend: Pushing firmly on crutches lifts uninvolved leg to step. Crutches and involved limb follow.

Descend: User places crutches on step below, pushing down on crutches to lift body weight; steps down with the involved leg followed by the uninvolved leg.

COMMON MEDICATIONS

Generic name	Trade name	Common use	Common side effects
Alprazolam	Xanax	antidepressant	sedation, possible tolerance
Amitriptyline	Elavil, Endep	antidepressant, decreases night muscle spasm	sedation, possible confusion
Aspirin	Many names	analgesic, anti-inflammatory, antipyretic	GI distress, blood thinning
Baclofen	Lioresal	antispasticity	transient drowsiness
Barbiturates	Phenobarbital (Luminal)	adult seizures	sedation, ataxia, skin problems
Benazepril	Lotensin	ACE inhibitor	headache, hypotension
Benzocaine	Americaine	local anesthetic	skin irritation
Carbamazepine	Tegretol	seizures	dizziness, drowsiness, ataxia, cardiac arrhythmias

Generic name	Trade name	Common use	Common side effects
Chlorzoxazone	Parafon Forte	muscle spasm	drowsiness, dizziness
Cimetidine	Tagamet	peptic ulcers	possible headache, dizziness
Dexamethansone	Decadron	corticosteroid	bone loss, muscle wasting, hypertension
Diclofenac	Voltaren	NSAID	GI disturbances
Diltiazem	Cardizem	Ca^+ channel blocker	headache, dizziness, hypotension
Etodolac	Lodine	NSAID	few
Fenoprofen	Nalfon	NSAID	GI side effects
Fluoxetine	Prozac	antidepressant	anxiety, nausea, insomnia
Ibuprofen	Motrin, Rufen	NSAID	some GI side effects
Indomethacin	Indocin	NSAID	some GI side effects
Lidocaine	Xylocaine	local anesthetic	some GI side effects
Nabumetone	Relafin	NSAID, analgesic	some GI side effects
Naproxen	Anaprox, Naprosyn	NSAID	some GI side effects
Phenylbutazone	Butazolidin	NSAID	some GI side effects
Piroxicam	Feldene	NSAID	some GI side effects
Prednisone	Deltasone	corticosteroid	bone loss, muscle wasting, hypertension
Procaine	Novocain	local anesthetic	nervousness
Rantidine	Zantac	peptic ulcers	possible headache, dizziness
Sulindac	Clinoril	NSAID	GI side effects
Tolmetin	Tolectin	NSAID	some GI side effects

APPENDIX

BIBLIOGRAPHY

General References

Grimsby, O. 1995. *Fundamentals of manual therapy: A course workbook.* San Diego, CA: The Sorlandets Institute.

Hoppenfeld, S. 1976. *Physical examination of the spine and extremities.* New York: Appleton-Century-Crofts.

Kessler, R.M., and D. Hertling. 1983. *Management of common musculoskeletal disorders.* Philadelphia: Harper & Row.

Kisner, C., and L.A. Colby. 1996. *Therapeutic exercise: Foundations and techniques.* 3rd ed. Philadelphia: Davis.

Kulund, D. 1982. *The injured athlete.* Toronto: Lippincott.

Magee, D.J. 1992. *Orthopedic physical assessment.* Philadelphia: Saunders.

Maitland, G.D. 1994. *Peripheral manipulation.* Boston: Butterworth.

Nordin, M., and V.H. Frankel. 1989. *Basic biomechanics of the musculoskeletal system.* Philadelphia: Lea & Febiger.

Norkin, C., and P. Levangie. 1992. *Joint structure and function.* Philadelphia: Davis.

Palmer, M.L., and M. Epler. 1990. *Clinical assessment procedures in physical therapy.* Philadelphia: Lippincott.

Paris, S. 1994. *Course notes.* St. Augustine, FL: Institute of Health Science.

Reid, D.C. 1992. *Sports injury: Assessment and rehabilitation.* New York: Churchill Livingstone.

Romanes, G.J. 1981. *Cunningham's textbook of anatomy.* New York: Oxford Press.

Roy, S., and R. Irvin. 1983. *Sports medicine: Prevention, evaluation, management, and rehabilitation.* Englewood Cliffs, NJ: Prentice Hall.

Soderberg, G.L. 1986. *Kinesiology: Application to pathological motion.* Baltimore: Williams & Wilkins.

Introduction

Cyriax, J. 1982. *Textbook of orthopaedic medicine: Diagnosis of soft tissue lesions.* Vol. 1. London: Bailliere Tindall.

Kaltenborn, F.M. 1980. *Mobilization of the extremity joints: Examination and basic treatment techniques.* Oslo: Olaf Norlis Bokhandel.

Kapandji, I.A. 1983. *The physiology of the joints.* New York: Churchill Livingstone.

MacConnaill, M.A., and J.V. Basmajian. 1977. *Muscles and movements: A basis for human kinesiology.* Baltimore: Williams & Wilkins.

Maitland, G.D. 1994. *Peripheral manipulation.* Boston: Butterworth.

Stockwell, R.A. Joints. 1981. *Cunningham's textbook of anatomy,* edited by G.J. Romanes. Oxford: Oxford University Press.

Williams, P., and R. Warwick, eds. 1980. *Gray's anatomy.* Philadelphia: Saunders.

Shoulder

Braatz, J., and P. Gogia. 1987. The mechanics of pitching. *The Journal of Orthopaedic and Sports Physical Therapy* 9: 56.

Butler, D.S. 1991. *Mobilisation of the nervous system.* New York: Churchill Livingstone.

Caillet, R. 1966. *Shoulder pain.* Philadelphia: Davis.

Carmichael, S.W., and D.L. Hart. 1985. Anatomy of the shoulder joint. *The Journal of Orthopaedic and Sports Physical Therapy* 16: 225-228.

Donatelli, R. 1987. *Physical therapy of the shoulder.* New York: Churchill Livingstone.

Donatelli, R., and M. Wooden. 1989. *Orthopaedic physical therapy.* New York: Churchill Livingstone.

Ferrari, D. 1990. Capsular ligaments of the shoulder: Anatomical and functional study of the anterior superior capsule. *The American Journal of Sports Medicine* 18: 20-24.

Fleisig, G.S., C.J. Dillman, and J.R. Andrews. 1991. A biomechanical description of the shoulder joint during pitching. *Sports Medicine Update* 6: 10-24.

Gowan, D., F.W. Jobe, J.E. Tibone, J. Perry, and D.R. Mangine. 1987. A comparative electromyographic analysis of the shoulder during pitching. *The American Journal of Sports Medicine* 15: 486-490.

Halbach, J.W., and R.T. Tank. 1997. The shoulder. In *Orthopedic and sports physical therapy,* 3rd ed., edited by T.R. Malone, T. McPoil, and A.J. Nitz. St. Louis: Mosby.

Hart, D.L., and S.W. Carmichael. 1985. Biomechanics of the shoulder. *The Journal of Orthopaedic and Sports Physical Therapy* 16: 229-278.

Hawkins, R.J., and J.C. Kennedy. 1980. Impingement syndrome in the absence of rotator cuff tear (stages 1 & 2). *Orthopedic Clinics of North America* 18: 151.

Jobe, F.W., and J.P. Bradley. 1989. The diagnosis of nonoperative treatment of shoulder injuries in athletes. *Office Practice of Sports Medicine* 8: 419-433.

Jobe, F.W., D.R. Moynes, J.E. Tibone, and J. Perry. 1984. An EMG analysis of the shoulder in pitching. *The American Journal of Sports Medicine* 12: 218-220.

Jobe, F.W., J.E. Tibone, J. Perry, and D.R. Moynes. 1983. An EMG analysis of the shoulder in throwing and pitching. *The American Journal of Sports Medicine* 11: 3-5.

Morrey, B.F. 1985. *The elbow and its disorders.* Philadelphia: Saunders.

Moynes, D.R., J. Perry, D.J. Anotnelli, and F.W. Jobe. 1986. Electromygography and motion analysis of the upper extremity in sports. *Physical Therapy* 66: 1905-1910.

Neer, C.S. 1983. Impingement lesions. *Clinical Orthopedics* 173: 70-77.

Pearl, M.L., J. Perry, L. Torburn, and L.H. Gordon. 1992. An electromyographic analysis of the shoulder during cones and planes of arm motion. *Clinical Orthopaedics and Related Research* 284: 116-127.

Peat, M. 1986. Functional anatomy of the shoulder complex. *Physical Therapy* 66: 170-179.

Perry, J. 1978. Normal upper extremity kinesiology. *Physical Therapy* 58: 265-269.

Poppen, M.K., and P.S. Walker. 1976. Normal and abnormal motion of the shoulder. *The Journal of Bone and Joint Surgery* 58-A: 195-200.

Schenkman, M., and V.R. DeCartaya. 1987. Kinesiology of the shoulder complex. *The Journal of Orthopaedic and Sports Physical Therapy* 8: 438-450.

Tank, R., and J. Halbach. 1982. Physical therapy evaluation of the shoulder complex in athletes. *The Journal of Orthopaedic and Sports Physical Therapy* 3: 108-119.

Thein, L.S. 1989. Impingement syndrome and its conservative management. *The Journal of Orthopaedic and Sports Physical Therapy* 11: 183-191.

Elbow

Andrews, J.R., and J.A. Whiteside. 1993. Common elbow problems in the athlete. *The Journal of Orthopaedic and Sports Physical Therapy* 17: 289-295.

Bowling, R.W., and P. Rockar. 1997. The elbow complex. In *Orthopedic and sports physical therapy,* 3rd ed., edited by T.R. Malone, T. McPoil, and A.J. Nitz. St. Louis: Mosby.

Morrey, B.F., L.J. Askew, K.N. An, and E.Y. Chao. 1981. A biomechanical study of normal and functional elbow motion. *The Journal of Bone and Joint Surgery* 63-A: 872-877.

Nirschl, R., and F. Pettrone. 1979. Tennis elbow. *Journal of Bone and Joint Surgery* 61-A: 835.

Regan, W.D., S.L. Korinek, B.F. Morrey, and K. An. 1991. Biomechanical study of ligaments around the elbow joint. *Clinical Orthopaedics and Related Research* 271: 170-179.

Stroyan, M., and K.E. Wilk. 1993. The functional anatomy of the elbow complex. *The Journal of Orthopaedic and Sports Physical Therapy* 17: 279-288.

Wrist and Hand

Conwell, E. 1970. Injuries to the wrist. *Clinical Symposia* 2: 3-30.

Craig, S.M. 1992. Anatomy of the joints of the fingers. *Hand Clinics* 8: 693-700.

Fisk, G.R. 1980. An overview of the injuries of the wrist. *Clinical Orthopaedics* 149: 37-143.

Jones, L.A. 1989. The assessment of hand function: A critical review of techniques. *The Journal of Hand Surgery* 14A: 221-228.

Kauer, J.M. 1980. Functional anatomy of the wrist. *Clinical Orthopaedics and Related Research* 149: 9-19.

Sarrafian, S.K., J.L. Melamed, and G.M. Goshgarian. 1977. Study of wrist motion in flexion and extension. *Clinical Orthopaedics and Related Research* 126: 153-159.

Volz, R.G., M. Lieb, and J. Benjamin. 1980. Biomechanics of the wrist. *Clinical Orthopaedics and Related Research* 149: 112-117.

Wadsworth, C.T. 1983. Clinical anatomy and mechanics of the wrist and hand. *The Journal of Orthopaedic and Sports Physical Therapy* 4: 206-216.

———. 1997. The wrist and hand. In *Orthopedic and sports physical therapy,* 3rd ed., edited by T.R. Malone, T. McPoil, and A.J. Nitz. St. Louis: Mosby.

Youm, Y., T.E. Gillespie, A.E. Flatt, and B.L. Sprague. 1978. Kinematic investigation of normal MCP joint. *Journal of Biomechanics* 11: 109-118.

Youm, Y., R.Y. McMurty, A.E. Flatt, and T.E. Gillespie. 1978. Kinematics of the wrist. *The Journal of Bone and Joint Surgery* 60-A: 423-432.

Hip

Crowninshield, F.D., R.C. Johnston, J.G. Andrews, and R.A. Brand. 1978. A biomechanical investigation of the human hip. *Journal of Biomechanics* 11: 75-85.

Dostal, W.F., and J.G. Andrews. 1981. A three-dimensional biomechanical model of hip musculature. *Journal of Biomechanics* 14: 803-812.

Harris, W.H. 1969. Traumatic arthritis of the hip after dislocation and acetabular fracture: Treatment by mold arthroplasty. *The Journal of Bone and Joint Surgery* 51: 737-755.

Kapandji, I.S. 1983. *The physiology of the joints.* New York: Churchill Livingstone.

Klampner, S.L., and A. Wissinger. 1972. Anterior slipping of the capital femoral epiphysis. *The Journal of Bone and Joint Surgery* 54-A: 1531-1537.

Radin, E.L. 1980. Biomechanics of the human hip. *Clinical Orthopaedics* 152: 28-34.

Saudek, C.E. 1997. The hip. In *Orthopedic and sports physical therapy*, 3rd ed., edited by T.R. Malone, T. McPoil, and A.J. Nitz. St. Louis: Mosby.

Knee

Cabaud, J.E., and D.B. Slocum. 1977. The diagnosis of chronic anterorotational rotary instability of the knee. *The American Journal of Sports Medicine* 5: 99-105.

Coplan, J.A. 1989. Rotational motion of the knee: A comparison of normal and pronating subjects. *The Journal of Orthopaedic and Sports Physical Therapy* 10: 366-369.

Frankel, V.H., and A.H. Burstein. 1970. *Orthopedic biomechanics*. Philadelphia: Lea & Febiger.

Fukubayashi, T., and J. Kurosawa. 1980. The contact area and pressure distribution pattern of the knee. A study of normal and osteoarthritic knee joints. *Acta Orthopaedica Scandinavica* 51: 871-879.

Goodfellow, J., D.S. Hungerford, and M. Sindel. 1976. Patellofemoral joint mechanics and pathology: 1. Functional anatomy of the patellofemoral joint. *The Journal of Bone and Joint Surgery* 58-B: 287-290.

LaFortune, M.A., P.R. Cavanaugh, H.J. Sommer, and A. Kalenak. 1992. Three-dimensional kinematics of the human knee during walking. *Journal of Biomechanics* 25: 347-357.

Nissel, R. 1985. Mechanics of the knee. A study of joint and muscle load with clinical application. *Journal of Biomechanics* 13: 375-381.

Paulos, L., F.R. Noyes, and M. Malek. 1980. A practical guide to the initial evaluation and treatment of knee ligament injuries. *The Journal of Trauma* 20: 498-506.

Timm, K.E. 1994. Knee. In *Clinical orthopaedic physical therapy*, edited by J.K. Richardson and Z.A. Iglarsh. Philadelphia: Saunders.

Wallace, L.A., R.E. Mangine, and T. Malone. 1997. The knee. In *Orthopedic and sports physical therapy*, 3rd ed., edited by T.R. Malone, T. McPoil, and A.J. Nitz. St. Louis: Mosby.

Foot and Ankle

Czerniecki, J.M. 1988. Foot and ankle biomechanics in walking and running. *American Journal of Physical Medicine and Rehabilitation* 67: 246-252.

DiStefano, B. 1981. Anatomy and biomechanics of the ankle and foot. *Athletic Training* 16: 43-47.

Donatelli, R. 1985. Normal biomechanics of the foot and ankle. *The Journal of Orthopaedic and Sports Physical Therapy* 7: 91-95.

———. 1987. Abnormal biomechanics of the foot and ankle. *The Journal of Orthopaedic and Sports Physical Therapy* 9: 11-15.

Hamilton, J.J., and L.K. Ziemer. 1981. Functional anatomy of the human ankle and foot. In *Proceedings of the AAOS symposium on the foot and ankle,* edited by R.H. Kiene. St. Louis: Mosby.

Hughes, L.Y. 1985. Biomechanical analysis of the foot and ankle for predisposition to developing stress fractures. *The Journal of Orthopaedic and Sports Physical Therapy* 7: 96-101.

Hunt, G.C. 1997. Examination of lower extremity dysfunction. In *Orthopedic and sports physical therapy,* 3rd ed., edited by T.R. Malone, T. McPoil, and A.J. Nitz. St. Louis: Mosby.

McPoil, T., and R.S. Brocato. 1997. The foot and ankle: Biomechanical evaluation and treatment. In *Orthopedic and sports physical therapy,* 3rd ed., edited by T.R. Malone, T. McPoil, and A.J. Nitz. St. Louis: Mosby.

McPoil, T., and H. Knecht. 1987. Biomechanics of the foot in walking: A functional approach. *The Journal of Orthopaedic and Sports Physical Therapy* 7: 69-72.

Murray, M.P., A.B. Drought, and R.C. Kory. 1964. Walking patterns of normal men. *Journal of Bone and Joint Surgery* 46-A: 335-360.

Nuber, G.W. 1988. Biomechanics of the foot and ankle during gait. *Clinics in Sports Medicine* 7: 1-13.

Seligson, D., J. Sassman, and M. Pope. 1980. Ankle instability: Evaluation of lateral ligaments. *American Journal of Sports Medicine* 8: 39.

Wright, D., S. DeSai, and W. Henderson. 1964. Action of the subtalar and ankle-joint complex during the stance phase of walking. *Journal of Bone and Joint Surgery* 46-A: 361-383.

Posture and Gait

Basmajian, J.V. 1979. *Muscles alive.* 4th ed. Baltimore: Williams & Wilkins.

Cavanaugh, P., and M. Laforune. 1980. Ground reaction forces in distance running. *Journal of Biomechanics* 13: 397-406.

Janda, V. 1983. On the concept of postural muscles and posture in man. *Australian Journal of Physiotherapy* 29: 83-85.

Kendall, F.P., E.K. McCreary, and P.G. Provance. 1993. *Muscles: Testing and function.* Baltimore: Williams & Wilkins.

MacKinnon, C.D., and D.W. Winter. 1993. Control of whole body balance in the frontal plane during human walking. *Journal of Biomechanics* 26: 633-644.

Mann, R.A., G.T. Moran, and S.E. Dougherty. 1986. Comparative electromyography of the lower extremity in jogging, running, and sprinting. *The American Journal of Sports Medicine* 14: 501-510.

McFadyen, B.J., and D.A. Winter. 1988. An integrated biomechanical analysis of normal stair ascent and descent. *Journal of Biomechanics* 21: 733-744.

Perry, J. 1992. *Gait analysis: Normal and pathological function.* Thorofare, NJ: SLACK.

Appendix

Ciccone, C.D. 1990. *Pharmacology in rehabilitation.* 2nd ed. Philadelphia: Davis.

INDEX

Note: Italicized numbers indicate illustrations; the *t* after a number refers to tabular information.

About the Authors

Janice Loudon, PhD, PT, SCS, ATC is an assistant professor in physical therapy education at the University of Kansas Medical Center and also provides clinical services at the medical center's Sports Medicine Institute. A sports certified specialist and certified athletic trainer, Dr. Loudon earned her PhD in movement science and biomechanics at Washington University in St. Louis, Missouri. In 1997, she completed a two-year program leading to certification in orthopedic manual therapy through the Ola Grimsby Institute. She is a member of the American Physical Therapy Association, the American Academy of Orthopedic Manual Physical Therapists, and the National Athletic Trainers' Association.

Stephania Bell, MS, PT is a clinical assistant professor in physical therapy education at the University of Kansas Medical Center. Bell teaches advanced orthopedics and provides clinical services at the medical center's Sports Medicine Institute. In 1997, she completed a two-year program leading to certification in orthopedic manual therapy through the Ola Grimsby Institute. Bell is a member of the American Academy of Orthopedic Manual Physical Therapists and the American Physical Therapy Association.

The administrator of two therapy clinics, one in Kansas City and one in rural Missouri, **Jane Johnston** holds an advanced masters degree in physical therapy from the University of Kansas and is a Certified Manual Therapist. Johnston is a member of the American Physical Therapy Association, American Back Society, American Academy of Orthopedic Manual Physical Therapists, and Women's Health Section, the Orthopedic Section, APTA.

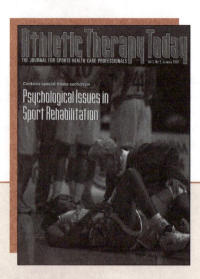
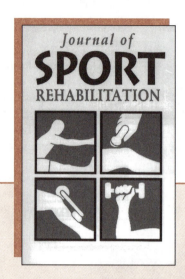